Pick-Your-Own Farming

PICK YOUR OWN FARMING

CASH CROPS FOR SMALL ACREAGES

RALPH L. WAMPLER
and
JAMES E. MOTES

UNIVERSITY OF OKLAHOMA PRESS

NORMAN

Library of Congress Cataloging-in-Publication Data

Wampler, Ralph L., 1929–
  Pick-your-own farming.

  Bibliography: p. 181
  Includes index.
  1. Farms, Small—United States. 2. Pick your own farms—United
States. I. Motes, James E. (James Earl), 1942–   . II. Title.
HD1476.U5W35 1984   631.'068   83–40333
ISBN 0–8061–1885–7

The paper in this book meets the guidelines for permanence and
durability of the Committee on Production Guidelines for Book Lon-
gevity of the Council of Library Resources, Inc.

Contents

v

Figures

Tables

Preface

There is virtually an untapped market for vegetables and small fruits grown close to towns, cities, and large metropolitan areas. Chain grocery stores buy a large part of their produce from California, Florida, and foreign countries. Transportation costs to ship that produce across the United States are increasing rapidly and are reflected in the price of produce in the stores. Meanwhile, some individuals in all the states are growing vegetables and small fruits and letting their customers pick their own produce at great savings over grocery store prices. All the harvesting labor is done by the customer. Only small acreages are needed for these tremendously profitable businesses, and though a Pick-Your-Own (PYO) operation is hard work in the summer months, the profits can be very worthwhile.

The PYO market goes begging in some states where customers must continue buying produce in chain stores at full and inflationary retail prices. Many of those customers would love to come to your place and gather their produce at large savings. Now is the time to get in on the profits.

One acre of a Pick-Your-Own operation can give you a good supplemental income. Ten acres can make you an excellent income. Forty acres possibly could make you rich. Chapter 2 in this book shows the profits that you can make growing vegetables and small fruits. Why not turn to that chapter and see how much money you can make per acre?

Pick-Your-Own Farming

1
Introduction

Literally thousands upon thousands of people in America today face financial crises heretofore unknown to them. Farmers are not exempt, because the price of everything that they use has soared, and the prices that they receive for their crops and stock have not risen in proportion.

Locating people to work on a farm is next to impossible, and when a farmer does find someone to work, wages must be competitive with industry or higher, because farm work lacks glamour and appeal.

Farmers have had to buy very expensive equipment that can be operated by one person. In many cases, they elect not to incur such large debts and just quit farming. Many farm widows face a difficult decision whether to continue operating their farm without their husbands or to sell out and move to town. Some farm people such as those move to the city. Some retire. Some get eight-to-five jobs and live unhappily ever after.

There are many people in cities from farming and rural backgrounds who would like to return to their farms, but they cannot because the farms have not produced enough revenue to support school-age children, buy a good automobile or boat, or whatever. Many of these people will spend 80,000 hours (a full working life) working at a job that they dislike. This book will show them how they may return to the farm as PYO operators and make more money than they now earn, working only a few months out of the year and working at a job that they prefer.

Many people reared in the city also yearn for a richer life in the country. They do not know how to live in the country, and if they move there, they can make mistake after mistake. This step-by-step guide to PYO farming will help them to avoid costly errors. It will place them almost on a par with the country folks at making a good income on a few acres.

Many people who are now retired on an acreage or farm are looking for a way to make extra money. This book is a natural for them. If they follow the procedures described here, these senior citizens can make more money than they ever thought possible. Similarly, senior citizens living in town who are able to plant a full backyard (approximately one quarter acre) can have a nice supplemental income.

There are many underpaid school teachers looking for a way to earn money in the summer months in some way other than manual labor that generates little more than the minimum wage. They should read this book very carefully. Hide it from your good teachers, however, because after one growing season they may quit teaching!

Is this book for everyone? Obviously not.

Pay no attention to what we say here if you live in the city and love your job or if you have no knowledge of country living and no curiosity about farm life. If you are a back-to-nature person or a homesteader, check this book only if you want to see how big money can be made on a few acres with minimal effort and time.

If your interest is just in enjoying country living, pass on—you do not need this book.

Now those of you who are interested in making money probably will not believe it when I tell you that you can *net* up to $4,000 per acre. Yet it can be done and is being done with regularity. Despite the high earnings only a few people are doing it because only a few know of the unbelievable potential. The secret is unveiled in this book step by step. You can do the same as those who are now making money literally by the bushel. The required investment is relatively small, the profits are great, and the risks are practically nil.

What is it all about, raising vegetables and small fruits for sale? People tend to think of a roadside fruit-and-vegetable stand with two barefooted children sitting in front looking hungry. In fact, a roadside stand is a very small, optional part of such a PYO operation. The pick-your-own marketing process is a far cry from the old fruit-and-vegetable roadside stand.

Read this book in its entirety before you decide to continue working year after year at a job you hate. In particular, read Chapter 2 on expected profits and let your eyes blink in disbelief. Tell everybody you want, they will not believe it! Old-time farmers will not believe it. Cattlemen will not believe it. But if you decide to go into this little-known business, do not tell your farm neighbors of your profits, because if they hear what you have done, they too will start small. Then after the first growing season, look out! All their efforts will be in the PYO area.

The PYO marketing method has been used successfully for years for three reasons: (1) city folks like fresh, high-quality vegetables and fruits; (2) by gathering their own, they make a sizeable savings over buying at the grocery store; and (3) for a city family a trip to a farm is not only a pleasant outing but also an educational trip for the children.[1]

Watch any neighbor of yours who starts in a PYO business and see if his economic level does not rise rapidly. Many people think that if you do not plant at least eighty acres, a farming operation is not worth the bother. That is bunk. Five acres of a good PYO business could very well net you more than 500 acres of wheat or another grain crop.

Take a look at the potential market. Vegetables and small fruits marketed directly to the consumer account for less than 3 percent of the total sold in the United States.[2] Included in that figure is produce sold at roadside stands, open-air markets, farmer's markets, and from trucks on the street, as well as PYO operations. Roadside stands received the greatest portion of the customers.[3] Although one can only guess, PYO operations probably account for considerably less than 1 percent of the vegetables and small fruits marketed in the United States.

People in Michigan, Pennsylvania, Ohio, Indiana, Illinois, New Jersey, and North Carolina have entered into PYO operations with some vigor. The PYO potential in the other states is staggering. Some states are virtually devoid of PYO operations. These virgin markets are waiting for you. Every settlement, community,

[1] The U.S. Department of Agriculture's 1978-79 report to Congress on the Farmer-to-Consumer Direct Marketing Program reported surveys revealing that 80 percent of shoppers patronized direct market outlets because of food quality while 60 percent preferred PYO operations because of savings.

[2] *Facts about Farmer-to-Consumer Direct Marketing,* Fact Sheet AMS-575, Agricultural Marketing Service, U.S. Department of Agriculture (August, 1978).

[3] U.S. Department of Agriculture 1978-79 report to Congress on the Farmer-to-Consumer Marketing Program.

town, city, and metropolitan area should be targeted. No matter how cold or how hot the climate is where you live, there is a potential PYO market. You may have to be selective about your crops, but the market is there.

The food in grocery stores, especially vegetables and small fruits, grows increasingly expensive because of growing transportation costs (caused by spiraling energy expenses) and the increasing cost of labor to harvest, pack, and store produce. In a PYO operation your harvest labor is supplied by your customers, and they pay you! Never before has the opportunity been greater to cash in on a PYO operation.

This book is directed at showing you how to make big money off a small acreage. Because large fruits, such as apples and peaches, require considerable acreage, they are not covered here.

Read this book with as much skepticism as you like. Then if you do not think that the procedures will work, try them on a small scale. They are working for others, and they will work for you. A good PYO operation *can* make you rich.

2

The Crops to Grow

Because this book is about how to make money in a PYO operation, the crops discussed here are the top money-makers. Others may be your favorites. If so, experiment with them by starting small. Then, if the crop proves to be a money-maker, grow it every year and expand the acreage to meet the demand.

At the beginning of your operation you should plant only the acreages that your customers will support, as suggested in table 1. If there is a PYO already in your area, find out what is planted there, because the market for some products may be saturated. The acreages in table 1 are based on the populations within twenty-five miles of your field. After the first season you will know how strong your market is for your particular crops, and you can make any necessary adjustments.

The profits that you make from your PYO operation will depend on many factors, but your management skills will be the primary factor. How you manage crop production to obtain high yields and quality is very important. Equally important is how well you manage the marketing of the produce. Getting the right number of customers to come to your farm when the produce is available will tax your skills. Undoubtedly, there will sometimes be wasted, unpicked produce on your farm, as well as unhappy customers unable to pick everything that they planned to buy on a particular visit. Your advertising program must be managed to minimize waste from unpicked produce and to maximize your profits and the number of satisfied customers.

Weather can ruin the best laid plans in a farming business. You will definitely have a problem if your strawberries or tomatoes are ripe and ready to pick and the weatherman promises rain. You must be flexible in these situations and alter your plans to minimize possible crop losses.

The profit from your PYO operation is the gross sales minus all the costs of growing your crops and running your operation. Gross sales are easy to determine by counting the money or checking the cash-register tape. Costs are much more difficult to determine. They can be separated into direct, variable costs and indirect, fixed costs. When the bills have been paid for seed, fertilizer, fuel, electricity, chemicals, labor, and other items, you will know how much the direct costs were. Indirect costs are more difficult to compute. They include items such as depreciation, maintenance on equipment and structures, taxes, insurance, interest, and payments for land. Once you are in business, many indirect costs will be incurred even if you do not grow crops.

The potential returns from the cultivation of one acre of selected small fruits and vegetables have been figured and arranged in tables 2 through 16 in this chapter. Table 2 is a summary listing of the crops, expected yields, selling prices, gross incomes per acre, direct growing costs, and returns per acre. The return figures are not the net profits because the indirect, fixed costs have not been deducted. The indirect costs were not deducted because they will vary greatly from one PYO operation to another. How elaborate an operation you have and the number of acres that you cultivate will greatly influence the indirect costs that will have to be prorated over each acre of production. For example, land costs will vary in different areas and locations. Investment in equipment and structures can be small or very large, depending upon your operation.

Tables 3 through 16 are in the alphabetical order of the names of the crops: asparagus, bell pepper, blackberry, blueberry, cabbage, cucumber, muskmelon, okra, snap bean, strawberry, summer squash, sweet corn, tomato, and watermelon. Broccoli and cauliflower are included with cabbage in table 7. The figures used in the calculations of the crop yields and selling prices are average values. Estimates for direct growing costs can be determined more precisely. In the individual crop tables the 1983 costs of standard soil-preparation operations and planting and growing practices have been used. Your actual direct growing costs will vary from those figures, but the information in the tables will help you prepare your production plans and arrange your financing.

THE CROPS TO GROW

Table 1.
Suggested Acreages Relative to
Population Within Twenty-five Miles of the PYO Farm

Crop	Population Within 25 Miles			
	10,000	50,000	100,000	500,000
Asparagus	1.0 acres	5.0 acres	10.0 acres	50.0 acres
Blackberry	1.0	5.0	10.0	50.0
Blueberry	1.0	5.0	10.0	50.0
Cabbage	0.1	0.5	1.0	5.0
Cucumber	0.1	0.5	1.0	5.0
Snap bean	1.0	5.0	10.0	50.0
Okra	0.1	0.5	1.0	5.0
Pepper	0.1	0.5	1.0	5.0
Squash	0.1	0.5	1.0	5.0
Strawberry	1.0	5.0	10.0	50.0
Sweet corn	5.0	25.0	50.0	500.0
Tomato	1.0	5.0	10.0	50.0

Table 2.
Estimated Costs and Returns per Acre of Selected Small Fruits and Vegetables

Crop	Yield (Pounds per Acre)	Price per Pound	Gross Income per Acre	Direct Growing Costs per Acre	Return per Acre (Gross Income Less Direct Growing Costs)
Asparagus	2,000	$0.80	1,600	$ 245	$1,355
Bell pepper	10,000	0.25	2,500	1,185	1,315
Blackberry	6,000	0.70	4,200	630	3,570
Blueberry	8,000	0.80	6,400	1,130	5,270
Cabbage	20,000	0.10	2,000	820	1,180
Cucumber	15,000	0.10	1,500	280	1,220
Muskmelon	14,000	0.12	1,680	525	1,155
Okra	8,000	0.30	2,400	242	2,158
Snap bean	6,000	0.30	1,800	310	1,490
Strawberry	7,500	0.50	3,750	840	2,910
Summer squash	18,000	0.15	2,700	320	2,380
Tomato	15,000	0.20	3,000	635	2,365
Watermelon	14,000	0.08	1,120	384	736
Sweet corn	1,200 dozen	0.75 dozen	900	330	570

Note: Indirect costs (such as mortgage payments and expenses for equipment and marketing) have not been deducted from the returns per acre in this and the other crop tables.

Table 3.
Potential Returns from an
Established One-Acre Planting of Asparagus

Gross income (2,000 pounds @ $0.80/pound)	$1,600
Direct growing costs	
Chopping fern	$ 15
Fertilizer	30
Weed control	30
Insect and disease control	20
Irrigation	50
Planting establishment cost*	100
Total	$ 245
Potential return	
(gross income less direct growing costs)	$1,355

*The planting establishment cost during the first two years is about $1,000 per acre, or $100 per year if prorated over ten years.

Table 4.
Potential Returns from a
One-Acre Planting of Bell Peppers

Gross income (10,000 pounds @ $0.25/pound)	$2,500
Direct growing costs	
Soil preparation for planting	$ 40
Transplants (10,000 @ $75/1,000)	750
Transplanting labor	75
Fertilizer	60
Weed control	60
Insect and disease control	100
Irrigation	100
Total	$1,185
Return (gross income less direct growing costs)	$1,315

Table 5.
Potential Return from an Established
One-Acre Planting of Blackberries

Gross income (6,000 pounds @ $0.70/pound)	$4,200
Direct growing costs	
Fertilizer	$ 50
Weed control	25
Insect and disease control	50
Irrigation	100
Pruning and training (30 hours @ $3.50/hour)	105
Planting establishment cost*	300
Total	$ 630
Return (gross income less direct growing costs)	$3,570

*Planting establishment cost during the first two years is about $3,000 per acre, or $300 per year if prorated over ten years.

Table 6.
Potential Return from an Established
One-Acre Planting of Blueberries

Gross income (8,000 pounds @ $0.80/pound)	$6,400
Direct growing costs	
Fertilizer	$ 50
Weed control	25
Replenish mulch	200
Insect and disease control	50
Irrigation	100
Pruning (30 hours @ $3.50/hour)	105
Planting establishment cost*	600
Total	$1,130
Return (gross income less direct growing costs)	$5,270

*Planting establishment cost during the first four years is about $6,000 per acre, or $600 per year if prorated over ten years.

Table 7.
Potential Return from a
One-Acre Planting of Cabbage

Gross income (20,000 pounds @ $0.10/pound)	$2,000
Direct growing costs	
Soil preparation for planting	$ 40
Transplants (14,000 @ $30/1,000)	420
Transplanting labor (30 hours @ $3.50)	105
Fertilizer	70
Weed control	40
Insect and disease control	75
Irrigation	70
Total	$ 820
Return (gross income less direct growing costs)	$1,180

Note: Direct growing costs for broccoli, cauliflower, and cabbage are similar. Broccoli and cauliflower yields are lower than cabbage yields, but broccoli and cauliflower bring higher prices. Thus the returns on broccoli and cauliflower should exceed the return for cabbage.

Table 8.
Potential Return from a
One-Acre Planting of Cucumbers

Gross income (15,000 pounds @ $0.10/pound)	$1,500
Direct growing costs	
Soil preparation for planting	$ 40
Seed (2 pounds @ $10/pound)	20
Planting	10
Fertilizer	60
Weed Control	40
Insect and disease control	50
Irrigation	60
Total	$ 280
Return (gross income less direct costs)	$1,220

Table 9.
Potential Return from a
One-Acre Planting of Muskmelon

Gross income (14,000 pounds @ $0.12/pound)	$1,680
Direct growing costs	
Soil preparation for planting	$ 40
Seed (2 pounds @ $10/pound)	20
Planting	10
Fertilizer	60
Weed control	40
Insect and disease control	50
Irrigation	60
Harvesting (70 hours of labor @ $3.50/hour)	245
Total	$ 525
Return (gross income less direct growing costs)	$1,155

Table 10.
Potential Return from a
One-Acre Planting of Okra

Gross income (8,000 pounds @ $0.30/pound)	$2,400
Direct growing costs	
Soil preparation for planting	$ 40
Seed (6 pounds @ $2/pound)	12
Planting	10
Fertilizer	50
Weed control	40
Insect and disease control	50
Irrigation	40
Total	$ 242
Return (gross income less direct growing costs)	$2,158

Table 11.
Potential Return from a One-Acre Planting of Snap Beans

Gross income (6,000 pounds @ $0.30/pound)	$1,800
Direct growing costs	
Soil preparation for planting	$ 40
Seed (70 pounds @ $1/pound)	70
Planting	10
Fertilizer	60
Weed control	40
Insect and disease control	40
Irrigation	50
Total	$ 310
Return (gross income less direct growing costs)	$1,490

Table 12.
Potential Return from an Established
One-Acre Planting of Strawberries

Gross income (7,500 pounds @ $0.50/pound)	$3,750
Direct growing costs (up to first harvest season)	
Soil preparation for planting	$ 40
Transplants (6,000 per acre @ $40/1,000)*	240
Transplanting labor (20 hours @ $3.50/hour)*	70
Fertilizer	60
Weed control	200
Insect and disease control	80
Irrigation	150
Total	$ 840
Return (gross income less direct growing costs)	$2,910

*Direct costs to maintain a bed for the second and third year would be lower because the cost of transplants and transplanting labor would not be incurred. The cost of narrowing the rows and thinning the plants after each subsequent harvest season would be about $100 per acre, however.

Table 13.
Potential Return from a
One-Acre Planting of Summer Squash

Gross income (18,000 pounds @ $0.15/pound)	$2,700
Direct growing costs	
Soil preparation for planting	$ 40
Seed (3 pounds @ $20/pound)	60
Planting	10
Fertilizer	60
Weed control	40
Insect and disease control	50
Irrigation	60
Total	$ 320
Return (gross income less direct growing costs)	$2,380

Table 14.
Potential Return from a
One-Acre Planting of Sweet Corn

Gross income (1,200 dozen ears @ $0.75/dozen)	$900
Direct growing costs	
Soil preparation for planting	$ 40
Seed (12 pounds @ $2.50/pound)	30
Planting	10
Fertilizer	60
Weed control	40
Insect and disease control	100
Irrigation	50
Total	$330
Return (gross income less direct growing costs)	$570

Table 15.
Potential Return from a One-Acre
Planting of Tomatoes (Not Staked or Pruned)

Gross income (15,000 pounds @ $0.20/pound)	$3,000
Direct growing costs	
Soil preparation for planting	$ 40
Transplants (3,000 per acre @ $75/1,000)	225
Transplanting labor	50
Fertilizer	60
Weed control	60
Insect and disease control	100
Irrigation	100
Total	$ 635
Return (gross income less direct growing costs)	$2,365

Table 16.
Potential Return from a One-Acre
Planting of Watermelon

Gross income (14,000 pounds @ $0.08/pound)	$1,120
Direct growing costs	
Soil preparation for planting	$ 40
Seed (2 pounds @ $12/pound)	24
Planting	10
Fertilizer	40
Weed control	40
Insect and disease control	40
Irrigation	50
Harvesting (40 hours @ $3.50)	140
Total	$ 384
Return (gross income less direct growing costs)	$ 736

3

Keys to Success

Although a good PYO operation can be a real money-maker, you should not read this book and think that by jumping into the PYO vegetable and small-fruit business you will be a successful farm entrepreneur tomorrow. Serious consideration should be given to several important personal and business matters.

YOUR APTITUDE

Before you go into PYO farming, you must determine whether or not you are the right sort of person to succeed in the retail farm business. You should look not only at the rosy money-making portions of this book. Basing their decision on that aspect of PYO farming, millions of people in our country would go into the business. Rather you must first decide without consideration of the profits whether retail farming is a lifestyle that you will enjoy. If you live or were reared on a farm, you know how that life is—it can be long hours of hard work in warm months. If you have never lived on a farm, you should spend time on one, helping the farmer, as an aid to making up your mind.

In addition to deciding whether you will like farm life, you must decide whether you are suited for retailing. If you have worked in a retail store of any sort, you know that a salesperson must deal with all kinds of people, some of whom will try your patience to the limit. Also, if you are unable to tolerate things that do not go according to plan, you had better stay out of the retail farm business.

Some farmers cannot stand to see food products that they have grown wasted. In a PYO operation you will see considerable quantities of produce ruined by customers. If your disposition will not tolerate the sight, you had better just stay with what you are doing.

On the other hand, if you have a great deal of patience and persistent enthusiasm and enjoy dealing with people and unusual situations, you are well-suited to PYO retailing.

TECHNICAL SKILLS

The skills required of a PYO operator are many. In addition to those required of any retailer, the PYO operator must know how to grow the products that he or she will sell. The PYO farmer needs to know how to prepare the soil; how to decide on the acreage for various crops; how to plant and cultivate the crops; how to apply fertilizer, herbicides, and insecticides; how to irrigate; and finally, how to retail the harvest to the public. He or she must know how to stagger plantings so that there will be a constant harvest. Therefore knowledge of the different plant varieties is essential. The varieties chosen must not only be resistant to disease but also early and late-bearing so that the harvest will last as long as possible.

If you have no farm background, you definitely should

work on a PYO farm before you make an irrevocable commitment to go into business yourself. Do not think you are a specialist because you have raised some tomatoes in your backyard. Growing vegetables on the magnitude required to make you rich requires real farm skills.

MANAGEMENT SKILLS

You possess considerable management skills, whether you are aware of it or not, if you now live on a farm. A farm wife, for example, generally is a good manager. She must plan and prepare meals and buy groceries, clothes, and many other things. She is able to have several activities going at one time, and with good planning, training and supervision things usually turn out right. Management comprises all such activities undertaken to accomplish goals.

Most people manage their household activities by routine or habit. You may be able to operate that way in a PYO operation with experience, but in the beginning while you are establishing new routines, you will need to be aware of what your goals are and what you will do to attain them.

This book sets out much of the planning required for a PYO operation. With a few modifications you can follow it for your operation. Excellent help is also available free of charge from your agricultural extension service. You should make use of that valuable resource. The extension service staff are eager to help and can advise you on a multitude of topics.

The better manager you are, the more money you will make in a PYO operation—and the more peace of mind you will have. The first step in good management is to determine your goal or objective. The goal may be stated

in dollars or in acreage or some other way, but no matter how you state it, write the goal on paper. See the
close of this chapter for more about drafting a plan.

LOCATION

Location is of critical importance. If you now own a
farm or acreage, or if you are planning to buy land for
a PYO operation, consider whether it is properly situ-
ated. Your property must be near potential customers
for your operation to be profitable. Surveys have shown
that about 55 percent of PYO customers come less than
twenty-five miles to buy their vegetables and fruits.[1]
About 24 percent come from twenty-five to fifty miles.
Almost 80 percent of PYO customers come a distance of
50 miles or less. The percentage of customers coming
over fifty miles is very small.

Gasoline costs have skyrocketed, and any savings
that customers incur by picking their own vegetables
may be offset by the increasing cost of gas. It may be
that because of the cost of gasoline your potential customers will no longer come as far as fifty miles. A safe
rule of thumb is that your property should be located
no farther than twenty-five miles from your customers.
Past studies indicated that transportation costs had
little bearing on the distances that customers traveled,
because they came together in one car, but it remains
to be seen whether that will be true in the future.

What size towns and cities do the customers come
from? Read and remember this: customers come from
small communities of less than 1,000 population; they
come from middle-sized communities; and they come

[1] Roger G. Ginder and Harold H. Hoecker, *Management of Pick-Your-
Own Marketing Operations* (Newark, Del.: University of Delaware Cooperative Extension Service, 1975).

from large metropolitan areas. One might think that people who live in small rural towns would not be interested in coming out to the edge of town and gathering strawberries, beans, and so on. Hogwash! These people have to buy groceries like everybody else, and all grocery stores buy the bulk of their food products from large wholesalers, who sell for approximately the same price to grocers in small towns as they do to grocers in large cities. There is a good market for your products, if you live within twenty-five miles of any town or city. How big that market is must be calculated. Obviously you do not need 1,000 acres of tomatoes for a town of 2,000. How you calculate the market potential of an area is set out in Chapter 4. Careful reading of table 1 at the beginning of that chapter will help you avoid underplanting or overplanting of a given crop.

ACCESS TO IRRIGATION WATER

Unless you are a most fortunate individual and live where rain falls every year in the right amount and at the right time for the different crops that you are planning to raise, you need to give serious consideration to the availability of water for irrigation purposes. Some of your crops will need much more water than others, but there usually are some dry years in which all crops require irrigation. The capability to irrigate when and where water is needed is critical for a sustained, money-making operation. If you rely solely on Mother Nature, you will have some good years, but you also may have years of total failure. The net results will be more serious than just not making money during a given dry year because some crops take as long as three years or more to become fully productive. An extremely dry year might kill the plants, and you would have to start the cycle over. For example, if you lost a crop of to-matoes, you would lose only one year's profits, because you would start over with new plants the next year anyway. On the other hand, crops such as strawberries and blueberries require a couple of years to become profitable.

DRAFTING A PLAN

To accomplish your goal, you must have a plan. Planning is extremely important to a PYO operation. A multitude of different activities should be covered, including the total acreage to be devoted to the operation, the acreage to be assigned to each of the crops, equipment needs and acquisition, money requirements, planting times, advertising methods and costs, personnel, and so on. Although most of the work will probably be done by you and members of your family at the beginning, your plan should include a timetable showing dates when tasks must be accomplished and by whom. Carry it through the harvest period.

The list goes on and on. The idea is to alert you to problems in the future. Although a plan is hard to write, it will be extremely important in attaining your goal. Write it out first in general terms and then fill in details as you think of them. After you have finished, follow the plan closely but be flexible enough to modify it when necessary.

People are not accustomed to writing their annual plans. Most plans are either made and kept in people's minds, or no detailed plans are made at all, except general plans such as "I want to put that forty acres down by the creek in hay this next spring." Writing a detailed plan is difficult and time-consuming, but it is worthwhile because your skills as a manager will be put to the test during your first season. You will have not only the crops to care for but also employee and customer problems. Your plan to accomplish your goal will determine your course of action. If you have planned properly when a new situation arises, you will probably not have to think it through and weigh the various possible responses, because you will have already done so and will have your decision in hand. You might say that your plan will determine standard operating procedures.

You will have plenty of novel questions to answer even with a good plan. For example, the best plan in the world will be no good if you are unable to manage people. Hiring, training, and supervising people will occupy much of your time. A plan gives you definite goals. At the end of the period covered you can sit down and see if you met the various goals and, if not, what went wrong that could be corrected. If you exceed your goals, you can consider to what you should attribute your success and whether it can be used again or used in another part of the operation to achieve even more. The following is a sample plan for a beginner PYO operation.

TINY YIELD'S PLAN FOR PYO OPERATION

Date
  August 4

Goal
  To make enough money from a PYO operation to go full time

Acreage Committed to PYO Operation
  Ten acres

Population
  Within 10-mile radius,   4,000
  Within 25-mile radius,   90,000
  Within 50-mile radius, 450,000

KEYS TO SUCCESS

Market Potential
Far greater than ten acres can satisfy

Competition
None except a peach orchard about 50 miles away

Acreage Division of Crops
1.0 acre tomatoes
0.1 acre peppers
1.0 acre okra
0.5 acre squash
1.0 acre snap beans
2.0 acres black-eyed peas
3.5 acres sweet corn
1.0 acre strawberries
Total   10.1 acres for PYO operation

Equipment Requirements
Except for a rototiller ($300) present equipment is sufficient for complete operation.

Irrigation Equipment
Surface pond is probably large enough for irrigation needs (used for years as flood irrigation for corn). Will continue this irrigation method except that on the one acre of strawberries will install sprinkler system to irrigate and to protect from late spring frosts.

Disposition of Existing Crops and Livestock
Determine whether more profitable to gather corn and fatten hogs or to sell hogs now and then gather corn and sell it separately.

Soil Preparation
October 1. Plow, disc, and plant rye as soon as corn is gathered. Includes the cow pasture in this if cows are sold. (It may take one or two years for this pasture to be ready for planting crops, but plowing and planting a good cover crop will help get rid of grasses, weeds, and insects.)

PICK-YOUR-OWN FARMING

Cash Requirements

Tomatoes	1.0 acre at $	635/acre = $	635
Peppers	.1 acre at	1,200/acre =	120
Okra	1.0 acre at	240/acre =	240
Squash	.5 acre at	320/acre =	160
Snap beans	1.0 acre at	310/acre =	310
Southern peas	2.0 acre at	250/acre =	500
Sweet corn	3.5 acre at	330/acre =	1,155
Strawberries	1.0 acre at	840/acre =	840

Total costs for crops* $3,960

Advertising	$ 100
Personnel (all labor done by family members first year)	–0–
Rototiller	300
Total cash requirement for first year	4,360

Cash on hand	$1,000
Proceeds from sale of hogs and cows	4,300
Proceeds from sale of corn	2,500
Total cash on hand	$7,800
Deficit	–0–

Soil Testing

Secure soil samples when plowing to test for lime and fertilizer needs. Take samples to extension agent to get soil test. At same time, advise him about the operation and see if there is a list of PYO operators for the state. If so, see about getting the operation listed. At the extension office pick up all vegetable production publications available from the agricultural college. This information will be needed to help learn the best production practices.

Time Table

August 15. Decision to sell hogs, cows, corn and devote efforts to PYO operation. Look for best buyer for each.

September 1. Disposal of livestock and corn. Start plowing and preparing soil for spring planting. Sow in rye for winter cover.

---

*This total includes the cost of soil preparation, chemicals, irrigation, and seed for each of the crops listed.

KEYS TO SUCCESS

September 15–March 1. Make road signs, see neighbors about putting up signs in good locations; prepare facilities, such as parking area with overflow area, and signs; acquire scales, containers, cash register, and any other equipment needed; then worry.

Before First Frost-free Date. Mark plots for the crops:

Crop	Plant Date*	Days After Planting Before Harvesting
Tomato	Frost-free date	70–120
Pepper	Frost-free date	70–120
Okra	2 weeks after frost-free date	60–first frost
Squash	Frost-free date	50–100
Snap Bean	1 week after frost-free date	60–80
Southern Pea	1 week after frost-free date	60–80
Sweet Corn	Frost-free date	70–80
Strawberry	2–4 weeks before frost-free date	during 3 weeks in late spring

End of Season. Disc and harrow soil and plant a winter cover crop, such as rye.

Conclusion

If Tiny can expand his operation to include his entire 40 acres, and when he gets his highly lucrative strawberries bearing, he will be able to attain his goal of working full-time on his farm— and make more money than he did before.

*See table 24 in Chapter 7.

Table 17.
Tiny Yield's Expected Income

Crop	Yield/Acre	Acres Planted	Price	Gross Income
Tomatoes	15,000 lbs.	1.0	20¢ lb.	$ 3,000
Peppers	10,000 lbs.	0.1	25¢ lb.	250
Okra	8,000 lbs.	1.0	30¢ lb.	2,400
Squash	18,000 lbs.	0.5	20¢ lb.	1,800
Snap beans	6,000 lbs.	1.0	30¢ lb.	1,800
Southern peas	3,000 lbs.	2.0	20¢ lb.	1,200
Sweet corn	1,200 doz.	3.5	75¢ doz.	3,150
Strawberries	No production until second year			
Total gross income				$13,600
Costs				−4,360
Income from PYO operation				$ 9,240

4

Getting Started

Nothing seems to go as fast as you would like when you start your operation. The key to success is following the directions in this book for growing the various vegetables and small fruits and growing the right amount of each. Use the formulas furnished in tables 18 through 20 at the end of this chapter. It does no good to grow ten acres of tomatoes if you only have a market for the produce from one acre. The other nine acres could be used to grow other highly marketable crops.

Your first year will probably not make you rich, because you will receive income only from growing and marketing those crops that will produce the first year. Some crops require as long as two or three years before they will bear a marketable harvest. Patience is highly rewarded in a PYO operation. If you are buying a small tract and leasing others, plant crops like blackberries, blueberries, and asparagus on your own tract because they have a long life. It takes three to four years to get blueberries into good production, but the planting will be fruitful for ten to twenty more years. If for some reason you do not buy the land you are leasing, you will still have your blueberries on your own land. You should plant crops such as tomatoes and green beans on leased land because the planting and harvesting is over in one season.

Start small in the PYO operation, whether you own your own land already or are now buying your land. You will gain valuable experience during your first season, and mistakes will not be so costly if your operation is smaller.

Another good reason for starting small is the per-acre cost of a PYO operation. A PYO operation is intensive in that you may make a considerable investment per acre compared to general farming. You want to make each acre produce as much as possible, which requires considerable expense in land preparation, planting, chemicals, and irrigation. Start small, and then the profits from the first season will finance an expanded operation the following year.

There are two people who can be extremely valuable to you if you are just starting. The first person is your spouse, if he or she is sold on the project. Show your wife or husband the money-making potential. If anything will convince a person, the information in tables 2 through 16, 19, and 20 will do so. You would have trouble hiring someone to do all the things a helpful spouse can do for you in a PYO operation. In fact, most PYO businesses seem to be family operations in which the spouse and children take active parts.

The second-most-valuable person to you will be your county extension agent. County extension services are free, and the agent can give you invaluable technical information on all phases of your operation. If you confront him or her with something he or she knows nothing about, the agent can contact an informed source to obtain the information. Get acquainted with your county agent before you start. He will certainly want to know of your operation.

WHAT SIZE FARM

If you now live on an acreage or farm, you will not have to face the same questions as the person who plans to buy land and start his PYO farm. What size farm you should buy depends, to a large degree, on what you want in life. If you want to make really big money, you would have trouble spending the profits made off of an eighty-acre PYO operation, if it is properly managed. About the smallest unit that will support you and a small family with the bare necessities is three or four acres. All of us can multiply the expected profits from a three-acre unit by multiples of ten and see pretty fast that a PYO operation is one of the most profitable farm operations today. When you consider the investment in money, time, and effort, a PYO operation is head and shoulders above wheat, corn, cattle, hogs, or what have you.

If you just want to get by and enjoy a leisurely pace, buy a small three- or four-acre tract and you have it made. If, however, you would like to get rich, but cannot afford a larger acreage, buy a smaller tract and lease other tracts with options to buy. If your desire is strong enough, you will figure out a way to own them.

After you have selected and bought your farm or acreage, you will have some idea about the acreage you will plant of the crops that you plan to grow. Table 1 gives you a good idea of how to allocate acreage. If there is a PYO operation already in your area, find out what is being grown there. The market for some products may be saturated.

ONE-ACRE SURPRISE

This book will tell you how to make big money on small farms and acreages. Perhaps you live in a town or city, and it is impossible for you to move to a location where you have access to more land, but you have a backyard, or perhaps an empty lot nearby, which can be planted in fruits and vegetables. Let us look at how much money one acre can generate if it is gardened intensively in vegetables. If you do not have access to one acre but only your own backyard, your income will be reduced proportionately. Let us place this one-acre farm in central Oklahoma, where there is a fairly long growing season. If you live farther north, the growing season will, of course, be shorter.

The major equipment items needed are a rototiller, spray can (unless you are growing organically), a hand-pushed garden seeder, and a good hoe and shovel. Irrigation can be done with a garden hose and lawn sprinkler. If you install a drip irrigation system instead of using the sprinkler, water usage will be cut in half.

Tables 18 through 20 are planting plans including spring, summer, and fall crops. Use of different areas of the one acre allows for crop rotation where the same crop is grown in both the spring and the fall season. The planting plan in figure 1 illustrates successive cropping to provide for intensive land use and maximum production from a small area approximately 209 feet square. In table 18 the needed seeds and transplants are given along with estimated costs. Other growing costs are also given in table 19 for this intensive one-acre operation. In table 20 are the crop yields by season and the income from one season's production for the intensively cropped acre.

Do you believe that you can gross $11,808 worth of vegetables from a one-acre plot as shown in table 20? The vegetable yields are reasonable to expect, and the sales prices shown are conservative PYO prices. The yield and sale prices are not inflated. The keys to the success of this one-acre project is the proper utilization

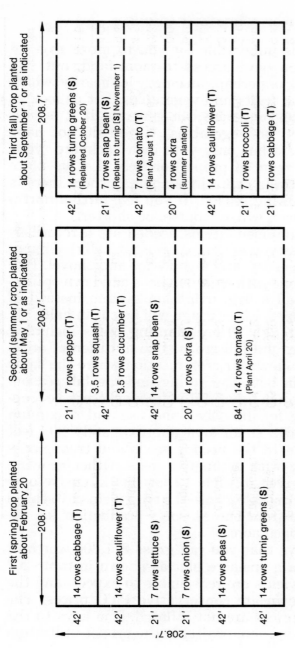

Fig. 1. Planting plan for intensive growing on one square acre of land in central Oklahoma.

First (spring) crop planted about February 20

◄─── 208.7' ───►

42' | 14 rows cabbage (**T**)

42' | 14 rows cauliflower (**T**)

21' | 7 rows lettuce (**S**)

21' | 7 rows onion (**S**)

42' | 14 rows peas (**S**)

42' | 14 rows turnip greens (**S**)

Second (summer) crop planted about May 1 or as indicated

◄─── 208.7' ───►

21' | 7 rows pepper (**T**)

42' | 3.5 rows squash (**T**)

42' | 3.5 rows cucumber (**T**)

42' | 14 rows snap bean (**S**)

20' | 4 rows okra (**S**)

84' | 14 rows tomato (**T**)
     (Plant April 20)

Third (fall) crop planted about September 1 or as indicated

◄─── 208.7' ───►

42' | 14 rows turnip greens (**S**)
     (Replanted October 20)

21' | 7 rows snap bean (**S**)
     (Replant to turnip [**S**] November 1)

42' | 7 rows tomato (**T**)
     (Plant August 1)

20' | 4 rows okra
     (summer planted)

42' | 14 rows cauliflower (**T**)

21' | 7 rows broccoli (**T**)

21' | 7 rows cabbage (**T**)

T, transplants (for cabbage, cauliflower, broccoli, squash, cucumber, tomato, and pepper).

S, crops seeded directly in the field (lettuce, okra, onion, pea, pepper, snap bean, and turnip).

Row spacing: 3 feet apart for cabbage, cauliflower, lettuce, onion, pea, pepper, snap bean, and turnip;
       5 feet apart for okra;
       6 feet apart for cucumber, squash, and tomato.

208.7'

Table 18.
Seed and Transplants for Intensive Growing on One Acre of Land

Crop	Acres Planted	Row Feet Planted	Seed Needed*	Transplants Needed	Price	Cost†
Snap bean	0.3	4,389	25 lbs.	—	$30.00	$ 30.00
Broccoli	0.1	1,463	1 oz.	1,463	0.05 ea.	73.15
Cabbage	0.3	4,389	2 oz.	4,389	0.05 ea.	219.45
Cauliflower	0.4	5,852	3 oz.	5,852	0.05 ea.	292.60
Cucumber	0.1	731	2 oz.	731	0.10 ea.	73.10
Lettuce	0.1	1,463	4 oz.	—	10.00	10.00
Okra	0.1	838	8 oz.	—	2.00	2.00
Onion (bunch)	0.1	1,463	8 oz.	—	8.00	8.00
Pea	0.2	2,926	20 lbs.	—	20.00	20.00
Pepper	0.1	1,463	1 oz.	1,463	0.10 ea.	146.30
Squash	0.1	731	2 oz.	245	0.10 ea.	24.50
Tomato	0.6	4,389	1 oz.	2,194	0.10 ea.	219.40
Turnip green	0.7	5,998	1.5 lbs.	—	4.00	4.00
Total	3.2			16,337		$1,122.50

*Seed needed for direct field seeding or transplant growing as appropriate for each crop.
†Costs include wholesale purchase of transplants and seed. A slightly lower transplant cost will be incurred if seed is purchased and transplants are grown in hotbeds or field beds.

Table 19.
Estimated Direct Cost of Intensive
Vegetable Growing on One Acre of Land

Item	Cost
Plowing and fitting for planting (custom-hired)	$ 100
Fertilizer	150
Seed and transplants	1,123
Seeding and transplanting labor	300
Weed-control chemicals and hoeing (includes $200 labor)	300
Irrigation (by drip system, including water cost)	500
Insect- and disease-control chemicals	300
Field cleanup between crops and rototilling labor	200
Total	$2,973
Cost of labor saved if you do your own work	−700
Total cost	$2,273

of all the area and putting in early and late crops as well as a crop at the usual garden planting time. The total cost in this one-acre project is $2,973, but $700 of that is for hand labor ($200 for weed control, $200 for cleanup between crops, and $300 for seeding and transplanting). If you do all your own garden work, you will have total expenses of $2,273, which will leave you a net of $9,535 ($11,808 − 2,273). If you have difficulty believing this, why not try it for one year? All the principles in this book about a PYO operation are valid on this one-acre project except that advertising is done by word-of-mouth, the cash register can be your pocket, parking can be your driveway. A small pair of scales would still be to your advantage, but otherwise the expenses of your operation are negligible.

Many people can make more money gardening one acre of vegetables than they would earn all year long on an 8:00-to-5:00 job. Think on it.

Table 20.

Crop Yields and Gross Income from Intensive Vegetable Growing on One Acre of Land

Crop	Spring Crop Yields (lbs.)	Summer Crop Yields (lbs.)	Fall Crop Yields (lbs.)	Total Crop Yields (lbs.)	Price (per lb.)	Gross Income
Snap bean	—	1,500	750	2,250	$0.30	$ 675
Broccoli	—	—	800	800	0.40	320
Cabbage	5,000	—	2,500	7,500	0.10	750
Cauliflower	4,400	—	4,400	8,800	0.30	2,640
Cucumber	—	2,500	—	2,500	0.10	250
Lettuce	2,400	—	—	2,400	0.20	480
Okra	—	1,000	—	1,000	0.30	300
Onion (bunch)	1,200	—	—	1,200	0.20	240
Pea	900	—	—	900	0.30	270
Pepper	—	1,250	—	1,250	0.25	313
Squash	—	2,800	—	2,800	0.15	420
Tomato	—	12,000	5,000	17,000	0.20	3,400
Turnip green	2,500	—	6,250	8,750	0.20	1,750
Total						$11,808
Less costs						−2,273
Net Profit						$ 9,535

5

The Picking Site

You should convey a natural image at the picking site. If you come from a farming background and want to convey a farmer image, it probably will be a success. If you cater to people from a large city, the farm appearance can be a real calling card. If, however, you wear overalls and try to talk like a farmer though you were reared in the city, you will likely be regarded as a fraud. Just be yourself.

If you have a small PYO operation, control at the picking site is not a big problem. If you have three acres or less in cultivation, you probably can control about everything at the picking site with little difficulty. But what if you have thirty acres in cultivation and customers are coming in by the dozens? Your operation, by necessity, must be very different.

PARKING

The first problem encountered in a sizeable operation is the question of where the customers will park. Do you let them decide? No. Help your customers by having the parking area clearly marked. The marking system does not have to be elaborate. For example, if you cut

39

a few pieces of timber and lay them where you want the cars stopped, that will be sufficient. Newly arrived customers will park pretty much the same way that vehicles are parked already, so set the example by parking your vehicle the way you want others to park. The parking area should be large enough so that cars can conveniently turn around and exit without bending fenders.

It is difficult to determine how much parking you will need. You want to use as little space as possible for parking because that part of your acreage is a total loss for growing crops. Most PYO operations need more parking than one might think, however, because the bulk of the customers come on the weekend. It is good psychology for your parking area to have a busy look about it. If a new customer drives into your parking lot and there are no cars, he or she may think something is wrong and that they are the only sucker around. They may just leave. Few things excite people in a PYO operation like a beehive of activity created by other customers. It seems to make people want to hurry and get into the fields to gather produce before everything is gone. At any rate, have as small a parking area as you can, so that no space is wasted and the lot will look busy to other customers. Then have an area adjoining the regular parking lot for overflow parking.

Here is a rule of thumb for the size of the parking area: you will need to allow space for ten cars in even the smallest operation. As the amount of acreage increases, allocate four parking spaces for every acre up to two hundred parking spaces. For example, if you have a ten-acre operation, you will need a minimum of forty parking spaces. Some farmers say that you need five to ten spaces per acre for orchards, and several times that number for strawberry fields.[1] A ten-car

[1] David W. Sams et al., *Pick Your Own Fruits and Vegetables* (Knox-

parking area takes about one acre after allowing for access aisles.[2] Costs will probably prohibit the surfacing of the parking area with gravel or other road material, but you should use an area that is at least well drained, covered with mowed grass, and free of mud holes.

CHECK-IN STATION

Customers should enter your premises at a central point where they can secure bags, boxes, baskets, and picking instructions. If customers furnish their own containers, the checker can weigh the containers as they enter the check-in station and mark the weights on the cartons or attach tags that tell the weights. Signs or handouts should inform customers of the prices of the various crops, the rules that they must obey, and any other information that you wish them to know. Signs should be kept to a minimum and should reflect an air of friendliness.

Customers should have a good feeling about coming to your place. The entrance should be clean and free of litter and trash. Some PYO operators like to greet customers personally and visit with them. If you like to do this, it will add to the appeal of your operation.

Not only should your parking and entrance area be inviting but also you should be careful about appearances. People will expect you and your employees to be dressed in work clothes, but they should be clean and presentable. Many customers have keen vision and a

ville, Tenn.: University of Tennessee Agricultural Extension Service, 1980).

[2]"U-Pick Farm Retailing Guidelines," a study adapted by James S. Toothman (Marketing Specialist, Pennsylvania State University Cooperative Extension Service) from a paper prepared by Ransom A. Blakeley.

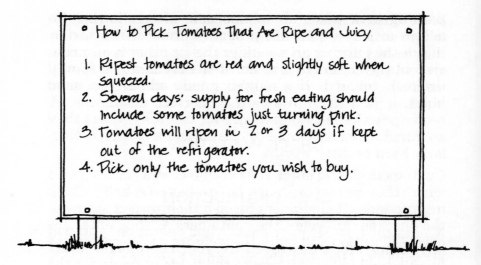

How to Pick Tomatoes That Are Ripe and Juicy

1. Ripest tomatoes are red and slightly soft when squeezed.
2. Several days' supply for fresh eating should include some tomatoes just turning pink.
3. Tomatoes will ripen in 2 or 3 days if kept out of the refrigerator.
4. Pick only the tomatoes you wish to buy.

Fig. 2. Necessary signs should be posted in fields where customers can see them or where the field supervisor can call attention to them. Instead of a sign, mimeographed information can be given to customers when they check in. The following information may be advertised on signs and in mimeographs: how to harvest crops; limitations on the size of the produce to be picked, the areas to pick, and so on; hours of operation; prices; the method of measurement for sale (by weight or volume); container information; minimum quantities, if any, that must be harvested; consumer information. *Drawing by Cynthia Marrs.*

keen sense of smell.[3] It is a delicate matter to have to tell an employee that his or her hair looks unwashed or that he has dirt or grime under the fingernails or a ring around the collar. It is easier if the checker is a family member, but if it is a neighbor lady whom you have hired, it is a little awkward to correct untidiness. A better practice is to look them over carefully before they are hired. Offer employees a cleanup break after particularly hard or dirty work.

CROP DESTRUCTION

No matter how large or small your operation, you will have some crops destroyed by your customers. Many families will bring children to the fields not only for the outing but also as a learning experience so that they will know how some of the foods they eat are grown. Some of these children will be in planted fields for the first time, and it is very likely that they will step on and ruin some produce. Although that is unfortunate, it is a loss that goes with this kind of retail farming. While some operators prohibit small children from going into the fields, many actually encourage customers to bring their children, and many of the children will help with the harvesting efforts. You must decide for yourself whether or not to let children in your field, and if you decide children are not to be permitted, by all means let that be known in your advertising and on your road signs. Parents often become angry when they learn at the check-in stand that children are not allowed in the fields. Some operators provide play areas for children

[3]Courtenay, Henry V., "What Makes People Buy?" *Consumer Economics* (Purdue University).

whose parents are in the fields. Insurance costs can be higher if play areas are made available for children.

A few parents will make no effort to stop the destruction of the crops by their children. Some children have no understanding of the value of the ripe vegetables and small fruits in the fields. They may very well throw produce at one another. Such actions should never be tolerated. Go immediately to the children and stop them. If they are unattended, take them to their parents, relate what you have observed, and make it plain that destroying crops is not permitted. Most parents will immediately control their children. If the same destruction is repeated by the same children, ask the whole family to leave the fields. Do not give a second warning—ask them to leave. A hospitable and friendly attitude is very important in a PYO operation, but customers should know that you will not tolerate willful destruction of crops.

MANAGING AREAS TO BE PICKED

Your plantings of a particular crop should be arranged so that those closest to the parking area and the check-in station are picked first. Thus customers are prevented from walking through plantings that are not yet ready for harvesting. After the first plantings are picked, you will, of course, want to direct your customers to other plantings that are ready for harvesting. A good way to keep customers out of areas not yet ready for harvest is to erect a barrier, such as a nylon string or engineer's tape strung on wooden or metal stakes. The barrier must be visible to customers. Some will cross over the barrier, but normally customers want to go where crops are ready to be harvested.

If you have a large operation, you will need signs

clearly marking the areas to be picked that day. Or better yet, have a field attendant point out the areas to be picked. The attendant should wear something that will distinguish him from the customers. During slack periods the attendant can pick various crops that you designate, so that they can be sold to customers who do not want to pick their own.

CHECK-OUT STATION

The check-out station should be noticeable even from a distance. You should place a large sign on the building, shed, trailer, or whatever you are using, stating that it is the check-out station. A large arrow should point to the exact place where the checker is located. In a small operation the check-out and check-in can be in the same station.

Do not make your customers wait in long lines when checking out. They will probably be tired and hot and have been in the fields longer than they originally intended, and a long, slow-moving check-out line can make people angry. The electricity bill for a long, slow check-out line in an air-conditioned building will make you mad. People want to get home fast when they come in from the fields, and you should do everything you can to speed up the check-out process.

Sometimes a line will form in spite of everything that you have done to expedite the check-out process. A great many customers may converge on the check-out station at one time, or perhaps a new and slow checker may be working. There should be a clear and obvious place for such a line to form. In a crowd of people attempting to form a line without guidance, the aggressive people will push the mild-mannered ones to the rear. Although you may not hear any complaints, this causes discontent and just plain anger. When a line begins to form, you should expedite the check-out process by opening another check-out line or by assisting the checker so that customers will wait only a short time. Some of your customers will want to visit with you. Do the visiting after the check-out process is completed so that other customers will not have to wait.

You will necessarily be thinking about who the customers are in a PYO operation. Most weekday customers will be women with children who are out picking crops to help the family food budget and because they seek top-quality food. The weekend customers will include some fathers who are not too happy about foraging in the countryside. Other husbands and fathers will jump into the spirit of the outing and have a good time. The weekends, especially Sundays, will be the busiest days.

Most customers are both health-conscious and money-conscious. The crops should speak for themselves in appearance, but you may be asked about the use of chemicals, and you should be completely honest and forthright in your answers. You should, of course, always follow the instructions on the pesticide labels. If you do not, you are violating the law and may cause people to become ill. If you have grown crops organically, this should be advertised, and customers can be directed in the direction of those crops, if they prefer.

Tell all new customers enthusiastically about your operation, especially during your first season. Try not to bore them, but let them know you try to grow fresh vegetables and fruits cheaper than they can be bought at grocery stores. That should be the main thrust of your talk. If you have the time and customers also want to talk about farm life, have at it. Those customers will be your advertisers.

But what about unreasonable customers who are just

Fig. 3. Check facilities should be as near the parking area as possible. They do not need to be elaborate, but they should be adequate for the number of customers expected. Clerks and checking equipment should be sufficient to keep customers from having to stand in long lines. A portable structure is useful to serve fields in different locations. *Drawing by Cynthia Marrs.*

47

not pleased with anything or anybody, including you? These people will not help your business, and you will be better off without them. Be cordial but firm. Tell them it appears they do not approve of much of anything you do and you will understand if they do not return. Thank them and tell them goodbye as you point toward the exit.

Check-Out Equipment

One piece of equipment that is a necessity in a PYO check-out stand is a good set of scales. Because speed is important, direct-reading platform scales are preferable to balancing scales, but anyway the scales you get should be positioned where they are level with the floor or a platform. Thus containers can be pulled over onto them, and produce will not have to be lifted. Even though you may sell some crops by volume or by counting the pieces, scales will be needed for most crops.

In addition to scales, you will need a good cash register that will produce a tape for the customer. It is desirable that the register also record the amounts on an inside tape for your protection and use. If you accept checks, use a rubber stamp with your name on it so that check processing will go faster. Also have a supply of ball-point pens available. If you charge sales tax, charts will save the checker much time and increase accuracy. If your prices are fairly stable, charts could be prepared for the checker for various crops.

The price of the various crops should also be marked and visible to customers as they enter and at the check-out area. A good chalkboard will suffice to display the various prices. The checker should also state verbally how the customer is being charged. For example, if tomatoes are being sold by weight, the checker should state the total weight and then tell the customer that

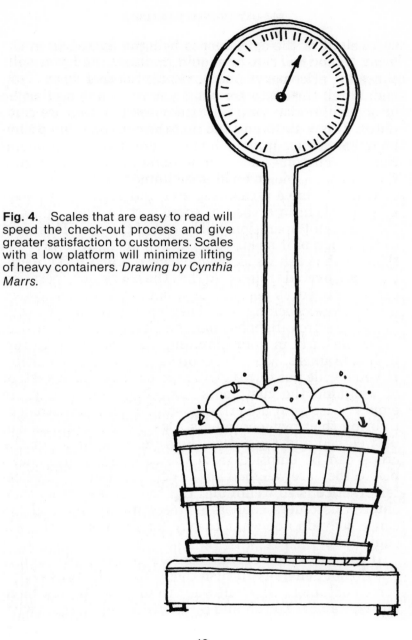

Fig. 4. Scales that are easy to read will speed the check-out process and give greater satisfaction to customers. Scales with a low platform will minimize lifting of heavy containers. *Drawing by Cynthia Marrs.*

the weight of the container is being deducted from the total. Then the checker should multiply the net weight times the price per pound out loud. In other words, you should tell the customer what you are doing so that he or she will understand how the various totals are computed. It is very important that the customer have complete confidence in your honesty and method of operation. If a customer does not understand how you figure the totals, suspicions will arise immediately.

You now have a cash register and scales. You need to place them in a shed of some sort so that a part of the check-out operation will be out of the sun. This shed or a portion of it needs to be locked at night to prevent theft of the cash register and scales. Some PYO operators use covered trailers, so that the check-out operation can be moved as needed. Customers may wish to rest a bit before checking out. They should be able to wait in a shady area. Benches under a shade tree will suffice.

Decide early in your planning whether or not to furnish containers. Most PYO operators do furnish them. If you decide not to, you will need to establish a system of weighing the customers' containers when they are empty. Also your advertising should tell the customers to bring their own containers. Customer containers can be weighed and tagged as they are brought in, or the crops can be emptied from the containers into sacks before they are weighed at the check-out area. The better practice is to weigh and tag the containers as they enter. Customers cooperate freely in this method because they do not want to pay for the weight of their containers.

It is best to have some containers available even if you do not generally furnish them. You will always have some customers who will want to pick but do not have any container. You should by all means equip them with

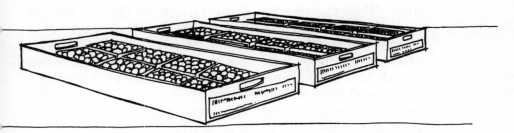

Fig. 5. It is simpler and more efficient to provide harvesting containers than to allow customers to bring their own. Uniform operator-supplied containers speed the checking process and also provide a place to print information such as the farm name, phone number, and location. *Drawing by Cynthia Marrs.*

containers, even if it is not your usual practice to do so, so that they can pick all that they want. You will also need to have plenty of strong paper sacks on hand, because over half of your customers will gather more than their containers will hold if you have good-quality crops. You will increase your sales significantly if you accommodate them with sacks.

If you do not know where to buy containers, check first in the yellow pages of your telephone book. If you find nothing there, check with competitors and see where they obtain containers. If you still have no luck, write the following companies:

Mesh Bags

Bemis Company, Inc.
800 Northstar Center
Minneapolis, Minn. 55402

Chase Bag Company
2 Greenwich Plaza
Greenwich, Conn.   06830

Nall Plastics, Inc.
108 West Second Street
Austin, Texas   78701

St. Regis Paper Company
150 East Forty-second Street
New York, N.Y.   10017

Wood Baskets

Berryville Basket Company, Inc.
Berryville, Va.   22611

F. D. Croce & Company, Inc.
One Mount Vernon Street
Ridgefield Park, N.J.   07660

Kingsville Basket Company, Inc.
3480 State Route 84E
Kingsville, Ohio   44048

Pacific States Box & Basket Company
Box 152
Glendale, Calif.   91209

Reds Package, Inc.
Route 9W
Milton, N.Y.   12547

Toilet facilities and fresh drinking water should also be available at the check-out stand. There are good portable toilets that would be excellent for the needs of your customers. The toilet should be kept clean at all times. Designate an employee to clean it and be sure that he knows it is his job and when it has to be done. Good drinking water is a must. Some customers will not be accustomed to working in the hot sun and will need a lot of water. Although a thirst can be quenched faster

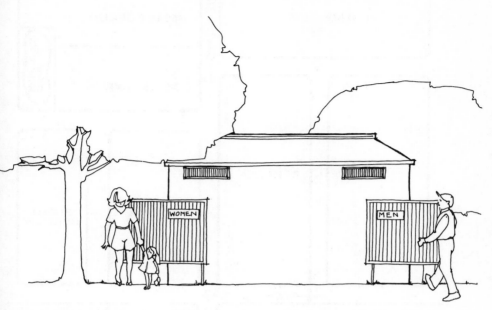

Fig. 6. Restrooms are necessities. They can be permanent facilities located near the parking facilities or they may be portable toilets positioned near the picking area or check stations. *Drawing by Cynthia Marrs.*

when drinking out of a container, paper cups are expensive and always make a place look messy. If you can, provide a fountain that has a drain for surplus water so that a mud hole is not created at the drinking area.

Cash register, scales, shed, toilet, and drinking water are minimum requirements for a good operation. What additional facilities or services you offer will depend on the size of your operation, what your competitors are doing, and your own preferences. Other possibilities are shaded tables for picnicking, food and drinks, ice, babysitting services, and a playground.

53

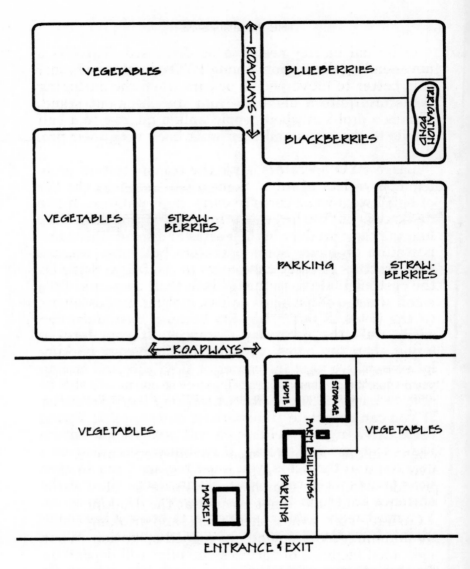

Fig. 7. Field layout should be determined by customer traffic flow and crop production requirements. Arrange plantings to maximize traffic flow and minimize crop damage by customers. For successful crop production you must consider your soils and topography, irrigation requirements, and rotations to reduce diseases. Use more elevated areas for plantings of fruits that are frost-sensitive. *Drawing by Cynthia Marrs.*

One final matter needs to be discussed. There is a never-ending discussion among PYO operators whether it is better to move people or cars when the gathering fields are quite a distance from the check-out stand. This is a problem when people walk a quarter to a half a mile to the crops and then walk back with their produce.

Large PYO operators solve the transportation problem in a variety of ways. Some offer customers the use of small wagons or carts to carry their produce. If the distance is not too far, this is much easier for them than lugging their produce in their arms and solves the transportation problem. Some operators pull large wagons with tractors to carry customers to the fields. Some let the customers drive to the fields in their own cars. On a small acreage of ten acres or less, getting the customers to the fields is not a big job because customers can easily walk the necessary distances. If your farm is larger than ten acres, consider making your parking lot somewhere near the center of the fields and having your check-out stand there. Thus customers will still be able to harvest crops without walking long distances. If you can eliminate transporting customers or letting them drive in your fields, you will prevent a lot of injuries to crops and people. If customers use their own cars to go to the fields, you must require them to open their trunks when they leave, placing large signs at the entrance to inform them that they will have to do so. You must require this of everyone, or you will hear complaints from those of whom you do require it.

Sell by Volume or Weight?

Selling your products by the volume of baskets, sacks, or boxes that you provide is easy and fast at

Fig. 8. Selling by weight solves the problem of customers overfilling containers and is fair to both buyer and seller. If you sell by weight, there is little problem in using containers of various sizes. *Drawing by Cynthia Marrs.*

the check-out stand. Yet this method can adversely affect your operation because of hurt feelings. For example, when is a bushel basket full of tomatoes? When the tomatoes are level with the top of the basket ring? When they are rounded a little on top? Or when the tomatoes are stacked as high as they can be without falling? You can get almost a bushel and a third of tomatoes in a bushel basket with a little effort. Then the bottom may fall out when the bushel is lifted. The same thing is true of other types of containers. The customer, of course, wants to get as much as possible for the money and loads the container until it can hold no more. The PYO operator wants to be reasonable and charge customers the same for the same amount of produce—and not reward good stackers. Of course, the PYO operator has legal control, but if he or she attempts to unload part of the good stacker's container, the customer becomes embarrassed, hurt, and mad. Such disappointed customers can kill much of the effectiveness of your advertising.

If at all feasible, you should avoid that sort of trouble by selling your products by weight. Selling by weight, however, will not work for all your products. Sweet corn, for example, is best sold by the dozen. If you have doubts about the best way to sell a particular product, try your idea. You can always return to the weight method. One PYO operator found that, when he sold bell peppers by the dozen instead of by the bushel, customers would pick the largest peppers and leave the small peppers that were not yet ready for picking. It was more trouble, but worth it.

EMPLOYEES

If you and your spouse or other partner are cultivating only three acres in your PYO operation, you probably will not need additional help except when sickness or business takes one away for awhile. By and large, the two of you should be able to operate the business easily.

If you move up to a thirty-acre operation, the picture is quite different. The most critical point in such an operation is where your customers enter and leave the property. Their containers must be weighed and tagged as they enter as well as checked out when they are finished picking. During peak periods you will need two very capable and personable people in the check-out area. At least one of them should be strong enough to lift, weigh, and sack produce. You must also have one person in the picking area to help direct customers and render assistance as needed. Whether or not you will be one of those three workers depends to a degree on your own desires. It is perhaps best for the manager not to be tied to a given spot and job. Rather you should be free to supervise the total operation and also to go to the wholesaler for sacks, to the bank with money, and so on.

Thus on a thirty-acre tract you need a total of five people: two checkers, a field attendant, your spouse or other partner, and you. You may think that you will not need two people at the check-out stand because you and your partner can handle that job. That may be true on slow days, but on busy days you may have to pull some vehicle out of a mud hole, go to the bank, or help with some of the many things that can go wrong. Your spouse may be too busy to work on the farm because of canning, cooking meals, cleaning house, or whatever. A working day on a PYO operation can be very long, and you are not in the business to ruin your health. Having help will enable you to do other things or just rest. During peak periods it may take four of you at the check-out stand to help customers. Also most employees will not work from daylight till dark, and you will need to be there when they are not on duty.

After you have decided on the number of employees needed, you will have to hire them. PYO operations provide excellent opportunities for retired individuals needing part-time work who are permitted to earn only a certain amount of money without affecting their Social Security benefits. Many such retirees still enjoy good health and make excellent employees. Many have spent time on farms or gardening and were schooled in an honest work ethic valuable to employers. In 1983, Social Security beneficiaries under age sixty-five could report earnings up to $4,920, and those over sixty-five could earn $6,600, without any loss of benefits.

College students home for the summer are also potential employees. Those from a farm background would be most valuable to you, but strange as it may seem, they are the least likely to work for you. In many cases they are trying to get away from farm activities. That is part of the reason why they are attending college. If they are planning to return to farming, they are planning something much bigger than your operation. Still, if you can hire a college student from a good farm background, he or she will probably make you a good hand.

High schools out for the summer are another source of employees, but be very selective in hiring because your employees will represent you with your customers. There are certain qualities that you should look for in prospective employees, no matter what their age group is.

Trustworthiness and Loyalty. These traits are at the top of the list of qualities that you should look for in prospective employees. These traits are especially important in employees who will be working the cash register. If you have any doubts whether a person can be trusted to operate your cash register honestly, do not hire him or her. There will be times when checkers will be working without you, and they must be trustworthy. If you think a prospective employee would short weigh pickings for a family member, do not hire that person. Employees should be told that the most important reason that they were hired was because you trust them at the cash register. Let them know that you trust them and they will more than likely honor that trust.

Dependability. Next after trustworthiness is dependability. An employee needs to have an established history and reputation of being dependable. Few things are more exasperating than to expect an employee at 8:00 a.m. and then have him arrive late time after time. Some tardiness is to be expected, but repeated tardiness should not be tolerated.

Competence. The next trait that you need to look for is competence. This normally will not present a big problem because most people today have sufficient education for PYO work, and with a little on-the-job training they can learn your operation readily. Of course, you should not hire a person who cannot tell a bell pepper from a green tomato. If farming is that foreign to him, his educational requirements for your operation are too great.

Personality. As noted previously, the atmosphere created by your employees will affect the image assigned to your operation by customers. Your employees need not be voted Mr. or Mrs. Personality of the community, but they certainly should not have personalities that would drive people away from your business. If the field attendant adopts the attitude that "they can find the field themselves," rest assured that your customers will

Fig. 9. Employees should be provided with a piece of easily distinguishable clothing that can be seen from a distance, such as a hat, shirt, apron, or armband. The farm name can be advertised on the clothing. *Drawing by Cynthia Marrs.*

lose some of their enthusiasm for patronizing your place. Probably the best advice you can give your employees is to be helpful and sincere. Sincerity will win over wit every time.

Dress. Special clothes should not be required of an employee, but employees should come dressed to work. They should wear khakis, jeans, or other suitable attire. Female employees will need to wear jeans or some other type of trousers; dresses will not do. As stated above, employees should also wear something to indicate that they are employees. It may be a special armband, tag, hat, or anything else that will distinguish them from customers. Then if customers need help or information, they know who to ask.

Hours. You may want to place limits on the number of hours that you will be open, and you may want to close some days. No matter what hours you are open or which days, there will be peak times when many customers will be at your place. You need help at those times.

Let us say, for example, that you have decided to stay open seven days each week from 8:00 a.m. to 6:00 p.m. Weekends will usually be your busiest times, especially Sundays. Customers will come on Saturday morning when you first open and start checking out about 10:00. a.m. There will be a slack period through the middle of the day and hot afternoon and then a flurry of activity again late in the day as temperatures cool down. The same thing will happen on Sunday except that Sunday morning is not as busy as Saturday morning, but activity late in the day is a little greater on Sunday than on Saturdays. So on weekends you will usually need help at the check-out stand from about 9:00 a.m. to 1:00 p.m. You will need help directing traffic in the fields the whole time because there may always be some customers needing help.

Perhaps you will want to take Mondays and Tuesdays off. You will need help for the whole day on those days. You and your spouse can handle the business alone for the rest of the week till Saturday. The field attendant may be needed during the week, but you should start off using him only on weekends. All the employees could be used more often if your business justifies it. For a thirty-acre operation you will need a minimum of two employees for four hours at the check-out stand on Saturday and Sunday, for a total of sixteen hours. One of those employees might work on Monday, and the other on Tuesday (your days off). Those would be ten-hour shifts for a total of twenty hours. The total number of hours of paid help at the check-out stand would then be thirty-six. The field attendant would be used all day Saturday and Sunday for a total of twenty hours, making the total number of hours for hired help fifty-six.

Orientation and Training. After completing the requisite pay forms, take time to properly orient and train new employees. Most new employees are willing to do a good job, but if their employer does not give them sufficient orientation, trouble arises. Conversely, time and money are saved if the employees are given detailed information concerning their responsibilities. Let them know what is expected of them.

If an employee will be working as a checker, tell the employee how it is done and then put him or her through a few dry runs. Make sure the employee knows how it is done before he or she starts processing customers. If customers are waiting to be checked out, let a new employee watch a few times and then work into the job little by little.

One part of the checker's job that is extremely important is making change to customers. Those of you who have worked cash registers may wonder why anyone would need help in this, but some of your checkers will be making change for the first time. It will save you money if you train them to make change correctly and efficiently.

The proper way to make change is first to take the bill given to you by the customer and place it on the table near you or on the cash register drawer, not in the cash register. Then get the correct change from the drawer and count it out loud to the customer. Only after the customer has indicated that the change is correct should you put the bill in the cash register.[4] What you are trying to avoid is a dispute over how large a bill the customer gave you. Trouble may arise if you place the bill in the register before counting out the change. If

---

[4]"U-Pick Farm Retailing Guidelines," a study adapted by James S. Toothman (Marketing Specialist, Pennsylvania State University Cooperative Extension Service) from a paper prepared by Ransom A. Blakeley.

you think you were given a ten-dollar bill, and the customer thinks it was a twenty, the argument begins. Such an argument is unnecessary and can be avoided if the instructions above are followed.

Sales on a particular day may amount to hundreds of dollars. Keep a close watch on the cash drawer and remove excess bills for safe keeping until you can go to the bank.

If you are required to collect any taxes (for example, a sales tax), prepare typed tables for the use of the checkers. The tables will speed up the check-out process and make collection of the tax much more accurate.

Be certain the employee who is working as the field attendant knows where the different crops are located. Explain where they are and why customers are not permitted in other areas. Tell the employee what questions customers are most likely to ask and also the appropriate answers.

Supervision. Your employees need some guidance, which we call *supervision.* As a general rule, you should supervise as little as possible to get the job done. Few things will make employees unhappy faster than too much supervision. When the boss looks over an employee's shoulder and corrects every little mistake, that employee will not be with the operation for long. People do not do things exactly alike, and employees will make mistakes. You should expect that. Also remember that, just because an employee does something differently, it does not mean that he or she is wrong. So give some leeway. If an employee does things detrimental to your business, by all means correct him immediately—that would be good supervision—but do not supervise just to show you are the boss.

Pay. How much an employee is paid is a matter of agreement between you and the employee, but it is most likely that you will be paying at least the minimum wage. An hourly wage competitive with industry wages is high, and not many employees are looking for an unglamorous job with a PYO operator. You probably will enjoy more success hiring retired people looking for part-time work or wives who have children and cannot work at a full-time job. Before you work out any pay agreement with an employee, check with the wage-and-hour division of the United States Department of Labor at the nearest federal facility. Also write your state department of labor to be sure that there will be no future complications.

Check with the Internal Revenue Service regarding withholding income and Social Security taxes. The IRS will provide the necessary forms and explanatory booklets on the taxes. When paying your employees by check, indicate on the check or on a separate slip, the gross amount earned and the deduction. The same information should be furnished if you pay in cash. This information removes any doubts from the employee's mind.

ROADSIDE STANDS

PYO operations were seldom heard of in earlier decades because farmers marketed their crops by setting up stands on the road near their farms or on a nearby highway where there was heavy traffic. Now most roadside stands have vanished. Most probably the stands required too much labor in harvesting and transportation. Farmers' children either are gone or have no interest in harvesting crops and sitting at a stand for long, hot hours.

The PYO operation eliminates most of the labor in connection with harvesting and transporting crops.

Fig. 10. A roadside stand or farm market can increase profits by supplying fresh-picked produce to customers who are not inclined to pick their own. Produce unavailable for PYO can be purchased and resold. Although some protection is needed for the produce, the structure does not have to be elaborate. *Drawing by Cynthia Marrs.*

Your customers perform both of those tasks, as well as the grading and packing. Yet occasions will arise when a customer wants to pick on shares. If your crops are plentiful, and you can sell the produce picked, by all means have a set standard for those who want to pick on shares and let them do so. Have a set standard for those who want to pick on shares. A standard share is two-fifths to the picker and three-fifths to you. If fifty-fifty sounds better to you, then use that division. If

there are competitors nearby, find out what split they
use. At any rate, you can sell your share to customers
who want fresh vegetables and fruits but do not want
to pick. Charge more for picked produce than for the
same crops in the fields. Charge about the same price
that the grocery stores are charging if your fruits and
vegetables are good or better than grocery store produce.

6

Business Necessities

ADVERTISING

The purpose of advertising is to draw attention to the existence and the superiority of your product. People must first know that your farm exists if they are to harvest your crops at the right time. You may have the best crops in the country, but if nobody knows to come and pick them, they will rot in the field. We all know that.

The idea is to have the maximum amount of advertising for the least cost. There are many different advertising media available to PYO operators:

- word of mouth
- newspapers *(including classified, space, and block ads and feature stories)*
- radio
- television
- direct mail
- telephone
- operator's sacks or other containers
- road signs
- chalkboards
- throwaway pamphlets
- novelties *(pencils, matches, caps sold by the PYO operator at a profit for sun protection)*

All of the above are effective advertising media, but you will want to use only those that give you the desired results at the least cost. You must bear in mind two questions: how much advertising you should use, and when you should use it.

First consider what type of advertising will bring you the most business. Surveys have shown that most customers learn about PYO operations from other customers, friends, and neighbors.[1] Such word-of-mouth advertising will work fine for you after you are in operation for a while, but what about during the first and second years when there are no satisfied customers to spread the word?

Classified newspaper advertising will reach the greatest number of potential customers for a given amount of money. When you place such an ad, you must be sure that people understand the what, when, and where of your operation. The following is a good model:

> Pick large, fresh strawberries [price]. Bring
> your own containers. Sunday, May—from 8
> a.m. till 6 p.m., 10 miles south—on Highway 4.

People pick their own vegetables and fruits primarily because of the quality of the fresh-picked produce and the low prices. From the advertisement above people will know that the strawberries are large and fresh— of good quality—and that they will save by picking the strawberries themselves. The ad also tells where the field is located. You can name your town if it is small, but if you are referring to a large city ten miles across, you had better use highway directions only, for example, "ten miles south of the intersection of Highways 10 and 4."

Classified ads are effective in small-town newspapers,

[1] Roger G. Ginder and Harold H. Hoecker, *Management of Pick-Your-Own-Marketing Operations* (Newark, Del.: Cooperative Extension Service, University of Delaware, 1975).

but your ad may be lost in pages and pages of ads in the classified sections of a newspaper serving a large city. Many shopping areas in large cities have a small, weekly tabloid that carries ads for the residents of the area. Those publications are good places for PYO ads.

Another effective way to advertise is to have a news story written about your operation. Newspaper reporters need and look for human-interest stories. If your operation is unusual in your community, you may be able to convince a reporter to visit your place for a story. Try to get a home-pages reporter if possible, because he or she will slant the story toward homemakers, who constitute the bulk of PYO customers. This kind of advertising is very desirable. It costs nothing, and the results are tremendous. If there are several PYO farms in your area, the newspaper probably will not be interested unless you offer a unique angle.

In addition to classified ads and human-interest stories, newspapers offer space advertising that is effective but somewhat expensive. If you feel that you need more advertising than a classified ad will give you, check the cost of space advertising in your local newspaper. Most newspapers will assist you in writing and designing a space advertisement.

Radio can be a very effective medium for advertising a good PYO operation. Your operation probably will not be large enough to support radio advertising for a while, but those who have been in the business for some time use it regularly. It seems that short announcements at regular times during the day work better than a long spot. Early morning radio ads are preferable because they do not compete with television. Many people listen to the car radio as they drive to work.

Television advertising can be effective, but usually only a large operation can support the costs. When you are that big, you will know which media serve you best. Most people think that picking their own vegetables and fruits in the field is much cheaper than buying the same crops in a grocery store. If they see your ads on television, they may think that, because television advertising is expensive, they will be paying for them and it might be just as cheap to buy at the grocery store.

Direct mail is an extremely effective advertising medium for existing PYO operators. The hard part is securing the mailing list. Most PYO operators let the customers fill out postcards addressed to themselves, checking the produce that they want to pick. The names and addresses on the cards are then recorded before they are mailed. Some PYO operators use a sign-in book to develop their mailing list. Other operators ask customers to complete questionnaires so that crops or services can be improved. With a place for names and addresses on such a questionnaire you can build your own mailing list.

The telephone is a good advertising tool, but it is used mainly to assure customers what crops are available. Many PYO operators record messages telling what crops are ready for picking. The reason for the recorded message is that hundreds of calls can come in on a given day, and it may take one person working full-time just to answer the phone. Some operators give a number to call at the end of the message if the person needs to talk to the owner about a particular problem. Customers may get the telephone number from any of the advertising media, but many times they take it from a container furnished or sold to them by the PYO operator when they last visited the farm.[2] The container

---

[2] David W. Sams et al., *Pick Your Own Fruits and Vegetables* (Knox-

or sack might also have a map drawn on it that would show a new customer the way to the farm.

Strategically placed road signs are effective and inexpensive advertising.[3] One difficulty is getting them located in high-traffic areas, such as near a well-traveled highway, because there are restrictions on the placement of such signs. You should check with local zoning authorities (usually in the nearest city) and with your state highway department. Then, if you find that you can get some road signs out, by all means do so. You should, in any case, mark your location with signs large enough to be read from a moving vehicle. The entrance and the parking area should both be clearly marked.

Inside the PYO operation a large chalkboard at the check-out place is an inexpensive advertising medium. Indicate on the board what crops are ready for picking, those that will be coming on, and the approximate dates that the harvests will start. Another homey advertising gimmick is to give away recipe booklets at the check-out stand. Not only do you furnish your customer with a good recipe for whatever he or she is currently picking but also you gain valuable advertising space on the backs of the pamphlets.

As we said at the beginning of this chapter, your most effective and least expensive advertising medium is word of mouth. For example, you should tell your county extension agent and your state department of agriculture about your operation, because people call them just to ask the whereabouts of your kind of business. In several states the extension service and the

---

ville, Tenn.: University of Tennessee Agricultural Extension Service, 1980).

[3] Antle, Glen, "Some Basics for Effective Signs," in *Proceedings 1978 Illinois Roadside Market Conference* (Urbana, Ill.: University of Urbana Department of Horticulture, 1978).

Fig. 11. An attractive sign listing the produce that you have available will reduce customer questions. It should be located near the check station or farm entrance. Keep the sign up to date to avoid disappointments and lost sales. *Drawing by Cynthia Marrs.*

agriculture department publish directories listing fruit-and-vegetable direct-marketing operations—you will certainly want your operation listed. One of the best places to spread the word is a beauty shop. Since women do the bulk of PYO, what better place to advertise by word of mouth than in a beauty shop? Take some of your produce to the owner of the shop and leave a few announcements telling what, when, and where you are operating. The word will be spread.

It is difficult to know how much advertising is enough until you gain some experience, but your first season will be very educational. One small word of caution:

do not overadvertise. In the first place, it is expensive. Secondly, your crops may be picked out early. It is very frustrating for a homemaker to make preparations to come to the fields with containers and to have all the children properly dressed for the occasion, only to find the crops all gone.

Be particularly careful about advertising strawberries and sweet corn. For some reason people get worked into a frenzy about those two crops. Those who have no particular interest in gathering other crops will drive miles to pick those two. Get ready for a crowd, when you advertise them, because it is going to happen.

Timing is the most difficult part of a PYO advertising program. Customers seem to come in bunches. What you would like is the exact same number of customers on each day that you are open. That would be the ideal situation. You would need a minimum of parking space and fewer employees, and thus your operation would be more profitable. Unfortunately, customers come to your place to satisfy their schedule not yours. They come when it is most convenient for them, usually weekends, though PYO operators have tried various methods. Properly planned advertising can only level out the peak loads of customers. For example, your advertising can indicate that the best picking is on Wednesday, or whatever day in the week you like. You might also give a discount for weekday customers and charge regular prices on the weekend.

No doubt you will be expert at advertising your business by the end of your first season, but do remember that good advertising does not have to be expensive to be effective. Reprinted below is "A Comparison of Alternative U-Pick Marketing Strategies for Strawberries," by Ransom A. Blakely.[4] Blakely compares various promotional strategies that may be used to gain profits from a ten-acre strawberry patch. You may, of course, experience different results, but Blakely is none-theless a recognized expert in PYO operations, and his conclusions are based on years of experience and study. Several things are noteworthy. First, Blakely provides good documentation of the folly of cutting prices. Price cutting does not increase sales as much as it produces customer dissatisfaction, because customers expect to buy the product at the same price for the remainder of the season. Second, Blakely reveals the value of radio and television advertising over newspaper ads when a message needs to be spread fast. Speed is of particular importance when weather disrupts the harvesting sched-ule of crops, such as strawberries, which deteriorate rapidly after ripening. Yet the most important message that the publication should relay to you is the profit that ten acres of strawberries can produce. Look at Strategy No. 4: the gross sales from a ten-acre straw-berry patch are $55,400! Another good feature of straw-berries is that the harvest period lasts for approximately four weeks.

Blakely's conclusions are as follows in his "Compar-ison of Alternative U-Pick Marketing Strategies for Strawberries":

INTRODUCTION

The following pages offer a comparison of six different strate-gies for marketing a crop of strawberries via U-pick. The data and assumptions, while hopefully within the realm of reality, were selected primarily for ease of calculation so the market-ing principles wouldn't become lost in the math. It would cer-tainly be advisable to substitute your own data or estimates and recalculate the possible consequences before adopting any of these strategies.

[4]This paper is available from the Department of Agricultural Eco-nomics, New York State College of Agriculture and Life Sciences, Cor-nell University.

ASSUMPTIONS

A 10-acre block of strawberries is expected to yield 10,000 pounds per acre this season. Maximum yield per day is estimated at 1,000 pounds per acre. To date, 20,000 pounds have been sold from this block at $0.60 per pound. Sales have been averaging 10 pounds per customer. The best day so far produced 400 customers and sales of $2,400.00.

Strategy 1: Sit on Hands

Sales next two days:

   1,000 customers x 10 lbs. each = 10,000 lbs. @ $0.60/lb. = $6,000.00

Weekend Return: $6,000.00

Berries remaining in field Saturday evening:

overripe	10,000 lbs.
still to ripen	60,000 lbs.

Strategy 2: Cut Price 10¢

Saturday sales:

   500 customers x 10.5 lbs. each = 5,250 lbs. @ $0.50/lb. = $2,625.00

   (For Saturday sales assume a slight increase in purchases per customer, from 10 lbs. each to 10.5 lbs., resulting from the price cut. For Sunday sales assume an additional 50 customers due to word-of-mouth promotion.)

Sunday sales:

   550 customers x 10.5 lbs. each = 5,775 lbs. @ $0.50/lb. = $2,887.50

Weekend Return: $5,512.50

Berries remaining in the field Sunday evening:

overripe	8,975 lbs.
still to ripen	60,000 lbs.

76

BUSINESS NECESSITIES

Conclusions

1. In spite of small increases in customer numbers and sales per customer, immediate returns are $488.00 less for Strategy 2 than Strategy 1.
2. There are now 8,975 pounds of overripe berries in the field, which may reduce the rate of pick of the remaining 60,000 pounds yet to ripen.
3. It will be difficult or impossible to bring the price back up to $0.60/lb. this season
4. The drastic price reduction may have aroused doubts in consumers' minds regarding the quality of the berries.
5. It will be difficult to increase the price next year to $0.65/lb. after customers have become accustomed to a $0.50/lb. price for most of the season.

Strategy 3: Cut Price 10¢ and Spend $500 on Advertising

(Assume the advertisements deliver 2,000 customers over the weekend.)

20,000 lbs. of berries ÷ 10.5 lbs. = 1,904 satisfied customers

2,000 customers at door — 1,904 supplied = 96 very unhappy customers

20,000 lbs. @ $0.50/lb. = $10,000.00

Weekend Return ($10,000.00 − $500.00 advertising cost): $ 9,500.00

Berries remaining in field Sunday morning:

overripe	0 lbs.
still to ripen	60,000 lbs.

Conclusions

1. In spite of spending $500 all in one day for advertising, net return for the weekend is $3,500 better for Strategy 3 than for Strategy 1.
2. There are no overripe berries remaining in the field to deter picking rate for the remaining 60,000 pounds.
3. The price has been spoiled for the remainder of this season.
4. Doubts regarding quality may have been aroused.
5. A price increase next year will be difficult.

Strategy 4: Hold Price and Spend $1,000 for Advertising

(Assume ads deliver 2,000 customers.)

2,000 customers x 10 lbs./customer = 20,000 lbs. sold

20,000 lbs. sold @ $0.60/lb. = $12,000.00 sales

Weekend Return ($12,000.00 sales − $1,000.00
advertising expenses): $11,000.00

Berries remaining in field Sunday evening:

overripe   0 lbs.
still to ripen   60,000 lbs.

Conclusions

1. In spite of spending $1,000 for advertising, immediate returns are $1,500.00 better for Strategy 4 than for Strategy 3 and $5,000 better than for Strategy 1.
2. All ripe berries were sold—future picking has not been harmed.
3. The $0.60 price can be maintained for the remainder of the season.
4. If customers attracted by the advertisements are happy with the quality of the berries, they are likely to return during the remainder of the season, helping to replace customers lost through normal attrition.
5. There should be no problem starting next season with a $0.60 or $0.65 price.

A CLOSER LOOK AT THE LONGER RUN

Strategy 1: Sit on Hands

Early-season pickings:

20,000 lbs. @ $0.60/lb. = $12,000.00 $12,000.00

10,000 lbs. @ $0.60/lb. = $ 6,000.00 $ 6,000.00

(Assume another 10,000 lbs. will be ready to harvest by Tuesday and that the 10,000 lbs. of overripes are still in the field. That means 50 percent of the berries will be overripe, or on average every other berry. A soggy and discouraging prospect for customers! So assume a picking rate of 20 percent for the remainder of the crop.)

Remaining season:

60,000 lbs. x .20 = 12,000 lbs. @ $0.60/lb. = $7,200.00 $ 7,200.00

Season Return $25,200.00

Strategy 2: Cut Price 10¢

Early-season pickings:

    20,000 lbs. @ $0.60/lb. = $12,000.00

Sales this weekend:

    11,025 lbs. @ $0.60/lb. = $ 5,512.50

By Tuesday there will be:

    8,975 lbs. of overripe + 10,000 lbs. of ripe = 18,975 lbs.

    (Thus 47 percent of the berries will be overripe—
hardly enough of an improvement over Strategy 1
to be noticed by consumers. Assume a picking rate
of 22 percent for the remainder of the crop.)

Strategy 3: Cut Price 10¢ and Spend $500 on Advertising

Early-season pickings:

    20,000 lbs. @ $0.60/lb. = $12,000 $12,000.00

Sales this weekend:

    20,000 lbs. @ $0.50/lb. − $500 = $9,500.00 $ 9,500.00

    (Because only ripe, ready-to-pick berries will be pre-
sented to customers Tuesday morning, assume a
picking rate of 90% for the remainder of the season.)

Remaining season:

    60,000 lbs. x .90 = 54,000 lbs.

    54,000 lbs. @ $0.50/lb. = $27,000.00 $27,000.00

Season Return $48,500.00

Strategy 4: Hold Price and Spend $1,000 for Advertising

(Assume advertisements deliver 2,000 customers.)

Early-season pickings:

20,000 lbs. @ $0.60 = $12,000.00 $12,000.00

Net sales this weekend:

20,000 lbs. @ $0.60 − $1,000 = $11,000.00 $11,000.00

(Since field was picked clean, assume a picking rate
of 90% for the remainder of the season.)

Remaining season:

60,000 lbs. x .90 = 54,000 lbs.

54,000 lbs. @ $0.60/lb. = $32,400.00 $32,400.00

Season Return $55,400.00

Conclusions

1. Price cutting alone may not move enough additional berries to maintain or improve net returns.
2. Effective advertising, by making consumers aware of product quality and availability, can be a more profitable strategy than price cutting. In a 1975 study, James Irwin found 57 percent of urban nonshoppers were not aware of the opportunity to obtain U-Pick crops near their home, yet over one-fifth of all nonshoppers owned a freezer and expressed an interest in trying U-Pick.
3. The importance of keeping the field picked clean so customers will continue to harvest a high percentage of the ripe berries is shown by the large differences in remaining season sales between Strategies 1 or 2 versus 3 or 4.

DISASTER!

The weather report was wrong! It rained all day Saturday! You had played Strategy 4, but no customers came. The most you can hope for is 1,000 customers on Sunday. Now what do you do?

Strategy 5: Turn 'Em Loose

Early-season pickings:

    20,000 lbs. @ $0.60/lb. = $12,000.00 $12,000.00

Sales this weekend:

    10,000 lbs. @ $0.60/lb. − $1,000.00 = $5,000.00 $ 5,000.00

Berries remaining in field Sunday evening:

    overripe   10,000 lbs.
    still to ripen   60,000 lbs.

      (Assume picking rate of 20%.)

Remaining season:

    60,000 lbs. x .20 = 12,000 lbs.

    12,000 lbs. @ $0.60/lb. = $7,200.00 $ 7,200.00

Season Return $24,200.00

BUSINESS NECESSITIES

Strategy 6: Strip Graze Half of the Field

Abandon half of the field—confine customers to the other half to obtain a clean pick.

Early-season pickings:

20,000 lbs. @ $0.60/lb. = $12,000.00 $12,000.00

Sales this weekend:

10,000 lbs. @ $0.60/lb. − $1,000.00 = $5,000.00 $ 5,000.00

Berries remaining in field Sunday evening:

	Picked Half	Abandoned Half
Overripe	0 lbs.	10,000 lbs.
Still to ripen	30,000 lbs.	30,000 lbs.

Assume 90% picking rate on clean-picked portion for remainder of the season.

Remaining season:

30,000 lbs. x .90 = 27,000 lbs.

27,000 lbs. @ $0.60/lb. = $16,200.00 $16,200.00

Season Return: $33,200.00

Conclusions

1. There are times when it may be more profitable to abandon a portion of the field in order to keep the remainder in suitable condition for customer picking, rather than attempt to keep up with the entire crop and suffer a poorer picking rate.
2. Another strategy would be to have hired pickers who can be called in on short notice to keep a field from going out of condition. By harvesting and selling or even dumping the overripes, they could

maintain the other half of the field in condition suitable for customer harvest.

IN SUMMARY

1. You should have a contingency plan. "Murphy's Law" says that "if something can go wrong, it will!" and with a U-pick operation, there is plenty of opportunity for things to go wrong.
2. An accurate and complete weather forecast is a most valuable tool to have in managing your harvest operation. If the forecast had called for showers Saturday and Sunday with intermittent clearing, then you might have been able to get some customers out in the fields if you had a good, well-drained sandy soil. So knowing not only that it is going to rain but how much and for how long can be very useful information in your planning.
3. A good estimate of the harvest in terms of the quantity expected to ripen each day can be most valuable in estimating how many customers you will need to clean up the field each day, thus how much advertising effort you will need and exactly when it will be needed.
4. You should have your advertising laid out in advance for different strategies. Newspapers don't give a great deal of flexibility. They require two to three days' lead time or more, but they can be useful for informing people about your berries, getting them to realize the importance of coming when the berries are at their peak, and getting them to listen to the radio station on which you can given them the day-by-day information.

There are more ways to kill a cat than by drowning him in cream. There are more ways to sell berries than by cutting the price. A bit of pencil pushing to compare the alternatives before the season begins could well save you from a panic decision during the season. The result would be more dollars in your pocket at the end of the season.

PRICING

It has been said that historically farmers do not set the prices for their products; they take the prices that are established by other people.[5] When they sell wheat, cattle, or whatever, farmers generally will sell for the prices that they are offered and do not bargain for a price in an arm's length transaction.

PYO operators are in a position to set their own prices for their products. Since few farmers know how to do this, one might ask what methods should PYO operators use to establish prices for their vegetables and fruits? We know that the prices that an operator asks should be less than the prices for the same products in local grocery stores. Few people will drive miles to your farm and go out in the hot fields to gather your vegetables and fruits if they can go to an air-conditioned store and buy the same products for the same price. On the other hand, people will come to your place because of the quality of your products and because they are less expensive than in stores.

Probably the factor that should most influence you in pricing your products is the prices charged by your competitors. You cannot charge 30 percent more than your competitors and expect a thriving business unless you have some other strong customer incentives offsetting the price (the quality of your produce might be such an offsetting factor). You should not, however, just check with your competitors and charge exactly the same prices that they charge. You should check their prices simply so that your price structure does not get out of line. You should never undercut a competitor's

[5] Roger G. Ginder and Harold H. Hoecker, *Management of Pick-Your-Own-Marketing Operations* (Newark, Del.: Cooperative Extension Service, University of Delaware, 1975).

prices as a way to secure more business.[6] If you do, he or she may then undercut you, and a price war will be on. Soon both of your operations will be history.

There are various methods of pricing retail goods. For example, you may add the cost plus a reasonable markup. To do that, you must, of course, know the real cost of producing your crops so that you do not charge less than what it costs you to produce them.

Odd as it may seem, the best pricing theory is to charge what the market will stand. For example, if your operation is located in an area where competition is practically nonexistent, you can charge a lot more than if the competition is fierce. If you have no competitors, you can determine the market price of vegetables and fruits by a fast stroll through your local grocery stores, noting the quality as well as the price of the various vegetables and fruits. You probably will find that you can sell your products in bulk at prices that are 10 to 15 percent lower than the retail price charged in grocery stores.[7] You can get rich with such a pricing system if you have enough acreage planted and practice good management. Check your grocery stores about once a week and also check newspapers for specials that most stores run when vegetables and fruits are in season.

[6] Blakeley, Ransom A., "The Paradox of Pricing for Direct Marketers," a paper presented at the United States Agricultural Marketing Service Workshop on Direct Marketing of Farm Products, Washington, D.C., 1979.

[7] Prices charged by PYO operators vary widely. In a paper presented at the Illinois 1978 Roadside Market Conference, J. W. Courter indicated that surveys revealed that operators were selling 10 to 20 percent below retail stores. Other papers indicate prices should be between the wholesale and the retail price. Prices for strawberries in Illinois in 1978 varied from 25¢ to 55¢ per pound. A multitude of factors may be the cause of this wide variance. See *Proceedings* of the 19th Annual Ohio Roadside Marketing Conference, held January 14-16, 1979 (Columbus, Ohio: Department of Agricultural Economics and Rural Sociology, Ohio State University), p. 121.

Most PYO operators believe that they should not raise the prices of their products during the season. If the customers expect a given price before they come out and then find the prices raised, they become very unhappy. Of course, it goes without saying that prices can be lowered and have great acceptance with customers.

RECORDS

Keeping accurate records is very important in a PYO operation. Of course, the first record you will have is the plan discussed in Chapter 3. It will be your road map to where you want to go and the various stops along the way.

You should keep other records as well: a diary, sales records, a payroll, and expense records.

Diary. The diary may be a book or a desk calendar with lines to write on. Keep it for as long as you are in business. Keeping an accurate PYO diary will place you ahead of your competitors. Any sort of information that might be useful for your next year's plan should be recorded in it. Information on the major happenings of the day should be entered. Describe the weather and record the sales each day. If you spray, put that in and state what kind of mixture you used. If a customer gets hurt, put that in, even if the injury was minor, and be sure to include the names and addresses of witnesses. That will be important information if the injured person decides to file a suit against you.

Sales Record. Your daily sales totals must be kept for tax reasons, as well as for your own information. You should record not only the total dollar sales each day but also the total sales for each of your major crops. For example, you may need or want to know the dollar sales from a two-acre tomato patch. If your cash register is sophisticated enough to record the sales of the separate crops, that is excellent. As a minimum your checkers could record in pounds or dollars the sales of any of the crops that you want to know about.

Payroll. You must keep payroll records so that you know the amounts of money you have paid to your employees and the amounts of Social Security and income taxes withheld. Periodically the proper amounts must be forwarded to the state and federal government agencies.

Expense Records. You will need records of your expenses to determine your net income for tax purposes. You also need these records for future planning. You can keep your finger on trouble areas in your operation if you have kept proper records. An expense record is one of the most important management tools.

INSURANCE

Whether or not you carry liability insurance is a personal choice. If you have a three-acre operation, you do not need as much insurance as the operator of a large unit does. For psychological and financial reasons, people are more likely to file suit against a large, solvent defendant than against the owner of a small operation, whose solvency may be in doubt. Yet people are generally much more litigious now than they were a generation ago, and some will file suit in response to the slightest provocation. It must be part of the get-something-for-nothing syndrome.

Many suits are filed for injuries received because of conditions that are only latently dangerous. Many more suits are filed by people injured because of conditions that are not so latently dangerous, known as "attractive nuisances." An example might be an old water well that had not been used for years, if a child played around it and fell in. You should take care that you have no such conditions on your property which are potentially dangerous to your customers or their children. Other examples might be animal traps; old windmills; machinery; animals, such as horses, dogs, and bulls; ponds; old buildings; and drifting chemicals. Look over your place and figure that if a person could be injured as it now exists, then he or she will be injured. Take corrective action to remove the risks.

The costs of insurance vary widely,[8] but at any rate the premiums are not expensive when computed on a per-acre basis, and especially when compared to the cost of a large judgment against you for some injury received in connection with your operation. Premiums vary not only from state to state but also with the type and size of the operation to be insured. The insurance agent will need to know the number of customers you expect, the sales volume, whether children will be permitted to pick, and the area involved—so that the company will know the hazards.

If your insurance agent tells you that you are covered, make him state that in writing. Insist that he mention in his letter that you are operating a PYO farm or that he issue you a policy stating that it covers your operation. The regular farm liability policy does not usually cover a PYO operation, because the risks were not contemplated when the policy was issued.

[8] A telephone call by one of the authors to a representative of a national insurance company revealed that the 1980 annual premium cost for straight liability up to $300,000 was from $100 to $150. The agent stated, however, that for what he called products liability (people getting sick from a residue of chemicals, for example) the annual premium was about $25 per $1,000 of sales. If you had sales of $50,000, your premium costs would be $1,250.

You probably have noticed by now that fruits grown on trees are not covered in this book, though apples and peaches have long been choice fruits for PYO operations. To pick them, you need ladders, benches, or some other means of getting the customer to the fruit. Although a ladder is indispensable around a farm, it is also the cause of many injuries. If you can make good money without exposing customers to such a hazard, why not do it? (Another reason for omitting orchards is that trees take up considerable space and the theme of this book is how to make money on small acreages. If you have a large farm and would like to raise apples and peaches, rest assured that the PYO market is strong.)

7

Successful Production Methods

IRRIGATION

Most crops require one to two inches of water each week from rainfall or irrigation. More water is needed by crops in drier areas of the country. Small plants generally require less water than larger plants. The volume of water required to cover one acre with one inch of water is 27,150 gallons. That is three or four times the amount of water used by the average family in a whole month.

Therefore considerable water must be available to irrigate a large area. Water sources include ponds, lakes, streams, and wells. A water source should be tested before it is used for irrigation. For a small fee state laboratories will determine if water quality is suitable for irrigating crops. Large lakes and streams are abundant water supplies, but wells are sometimes limited. A rule to follow is that a well needs to supply ten gallons of water per minute for each acre it irrigates. A well should have the capacity to supply one hundred gallons per minute if ten acres are to be irrigated.

Irrigation is used to supply water to crops during the growing season when rainfall does not provide enough water. Even where average rainfall usually provides enough water, it seldom comes just as needed during a growing season. Irrigation is essential to guarantee good yields and high-quality crops for PYO. The need will vary greatly from year to year in humid regions, depending upon the amount and distribution of rainfall during the growing season, but you will still benefit from irrigation even if your area has high rainfall.

Several methods are used to irrigate crops. Before deciding which one is right for your operation, you must consider what kind of soil you have, how much you can invest, the amount of water available, and the labor available to do the irrigating.

Surface Irrigation. Running water over the surface of the soil is the cheapest way to irrigate. Surface irrigation can be used only on level or near level land where water is available at low cost. Two methods of surface irrigation are used: furrow and flood. In furrow irrigation water is run through furrows between the crop rows or beds. In flood irrigation the area to be watered is surrounded by a dike and water is allowed to flow over the field. The main advantage of surface irrigation is its low cost; it requires little equipment.

Disadvantages of surface irrigation include uneven water distribution and water loss in sandy, porous soil where the water soaks in too fast and does not cover the field. In such terrain labor with shovel in hand is needed to make sure all parts of the field are being watered.

Sprinkler Irrigation. Sprinkler irrigation is commonly used by vegetable growers in many areas. The many different sprinkler systems vary greatly in cost and in the labor required to operate them. All require a large pump to pressurize the water. About forty to fifty pounds of pressure are required to break water into small drops and evenly cover the soil.

A simple sprinkler system can be a single irrigation line consisting of joints of pipe that reach from one end of the field to the other. Each section, or joint, of pipe has a sprinkler that rotates and waters a circular area. The wetted areas from neighboring sprinklers overlap so that generally a strip about sixty feet is wetted. Pipe without sprinklers connects the pump to the irrigation line. It is important that the water source be fairly close to the field to be irrigated.

When enough water has been sprinkled on one strip, the entire line is moved forty to sixty feet to one side, and the adjoining land is irrigated. Although this system is fairly inexpensive, considerable labor is needed to move the entire line of pipe when a new area needs watering. Although it is made from aluminum, a 30- or 40-foot joint still weighs around fifty pounds. That may not sound heavy, but moving sprinkler irrigation pipe is not fun when you are sinking into mud halfway to your knees—and remember, some of those joints will contain water.

Little labor is required to irrigate with a solid-set system once the system is in place, but having the irrigation pipe laid out all over the field can be a headache when the crops need to be cultivated or sprayed. Some of the pipe will have to be moved, but it usually only needs to be moved a short distance, and the soil will not be muddy during cultivating or spraying operations. The initial cost to have an entire field covered by solid-set irrigation pipe and sprinklers is around $1,500 per acre. Some crops, such as strawberries, will justify the expense. (For strawberries a solid-set system is used to supply water not only for growth but also to prevent late spring frosts from killing early flowers).

Two other kinds of sprinkler irrigation systems have been developed in recent years: center-pivot systems and traveler or "big-gun" systems. In a center-pivot system a single line of pipe with sprinklers extends from the center of the field to the outer edge. The long line of pipe is supported by large towers mounted on wheels. It rotates around the center point, covering the field with irrigation water. Field shape is important: center-pivot systems work better in square fields than in rectangular ones, but even in a square field an area in each corner cannot be watered. Although very little labor is required, a center-pivot system is a poor choice for a smaller PYO operation because fields must be forty acres or larger for the system to be practical. Also there is less flexibility in center-pivot systems to water a particular portion of a field, as is necessary in a PYO operation where several different crops are grown.

In a traveler or big-gun system, water is pumped through one large sprinkler at high pressure, and a single sprinkler or gun waters a strip 200 to 300 feet wide. A flexible hose feeds water to the gun, which moves through the field slowly, powered by a winch. A traveler system costs much less than a center-pivot system, but labor is needed to set the system up for each pass through the field. Traveler systems come in various sizes, and a smaller size works in most PYO operations.

A third type of sprinkler is the side-roll irrigation system. The side-roll system is similar to a hand-moved line except that the entire irrigation line is on wheels, with the irrigation pipe as the axle. Only a short time is required for one man to move the entire line of pipe because an engine rolls the line from one irrigated area to another. An entire field is covered in a series of moves of about sixty feet each. Side-roll systems can be as much as one quarter mile long. Fields need to be square or rectangular in shape. Since there is only about three feet of clearance under the irrigation pipe, side-roll systems cannot be used in tall crops, such as sweet corn and staked tomatoes. Side-roll systems are good to irrigate fifteen or twenty rows of a particular short-growing crop and quickly move over crops not needing water to other areas of the field.

Each individual sprinkler irrigation system has advantages and disadvantages. In general, the advantage of sprinkler irrigation compared to flood irrigation is that it can be used on rolling, uneven land and on sandy, porous soils. Sprinkler irrigation distributes water fairly uniformly, and the rate of water application can be adjusted to suit the soil. Light applications of water can be made on shallow-rooted crops; water is not wasted by soaking in the soil below the crop roots. See the relative rooting depths of crops in table 21.

There are several disadvantages to sprinkler irrigation systems, compared to flooding methods. Both initial and operating costs are higher because a larger pump and motor are needed. Moving the pipe or the whole sprinkler system requires labor, and because moving pipelines is very undesirable work, your labor may become discontented. Strong winds can blow the water and cause uneven water distribution. Sprinkler nozzles can clog, requiring frequent attention.

Drip Irrigation. Drip irrigation is a relatively new irrigation method that has been used by vegetable and fruit growers for the past ten to fifteen years. A drip system uses a small hose to supply water to each crop row and each plant. It differs little from the old garden soaker hose except that durable, low-cost drip lines have been developed for use on large acreages. The irrigation water

Table 21.
Relative Rooting Depth of Crops

Shallow (1 to 2 feet)	Medium (3 to 4 feet)	Deep (4 or more feet)
Broccoli	Bean	Asparagus
Blueberry	Beet	Pumpkin
Cabbage	Blackberry	Sweet potato
Cauliflower	Carrot	Tomato
Lettuce	Cucumber	Watermelon
Onion	Eggplant	Winter squash
Potato	Muskmelon	
Radish	Pea	
Spinach	Pepper	
Strawberry	Summer squash	
Sweet corn	Turnip	

Note: Adapted from O. A. Lorenz and D. N. Maynard, *Knott's Handbook for Vegetable Growers* (New York: John Wiley and Sons, 1980).

must be clean—even filtered where necessary. A drip system is versatile and can be put together to irrigate any row spacing or row length. It eliminates spraying water with sprinklers or running water down furrows. The distribution of water is very uniform and unaffected by wind.

Lower irrigation costs are the main advantage of drip irrigation, which uses less than half the water required for furrow or sprinkler irrigation. Labor costs also are lower for drip irrigation: because the water is applied slowly, line and pipe sizes can be smaller; because less water pressure is needed, smaller pumps and motors are adequate. Because much of the soil surface does not become wet while irrigating, customers can enter the fields

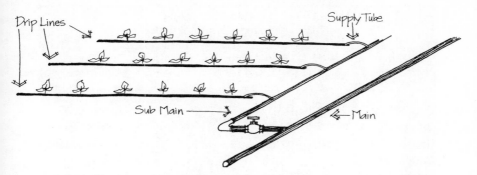

Fig. 12. The drip lines of a drip irrigation system supply water to plants in the rows. *Drawing by Cynthia Marrs.*

sooner, and weed control is easier. Because soil moisture is more evenly maintained, crops grow better.

First among the disadvantages of drip irrigation is the need for clean water. Usually filters are needed. Also, drip lines must be removed after harvest before the field can be tilled. Fortunately, most drip lines last several years, and they can be rolled up and stored for use the next season.

FIELD EQUIPMENT

Several farm-equipment items are needed to grow small fruits and vegetables for PYO. For an operation of thirty acres or less owning a large horsepower tractor is not quite economical. Hire your neighbor to do big jobs like plowing and discing. A 25- to 35-horsepower tractor has enough power for most other PYO operations. It can level soil for planting and pull the other pieces of equipment that you will use. If your plowing and discing are done by a neighbor, you will need only the following equipment items:

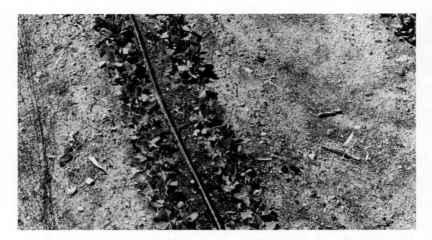

Fig. 13. For drip irrigation a drip tube is placed along each row, or a single tube is placed between two closely spaced rows, as shown here. Drip irrigation continuously wets a strip in each row. Other systems release water only in the immediate area of trees and bushes in small-fruit and orchard plantings.

Fig. 14. Drip irrigation lines can be installed under plastic or other mulching materials, or buried below the soil surface. Here a "lay-flat" main line is supplying water to drip lines under plastic mulch. A plastic-mulch laying machine can be modified so that drip tubing and mulch are laid in the same operation.

- 25-35 horsepower tractor
- harrow (to level disced soil before dragging)
- drag (to put final touches on a seedbed before planting)
- fertilizer applicator
- planter
- transplanter
- sprayer (for pesticide application)
- cultivator (for weed control)
- pickup truck or trailer (to haul supplies)
- hoes, rakes, and shovels (for hand weeding and thinning crops)

Most of the above items can be purchased used at farm auctions or farm machinery stores. Used equipment does just as good a job as new equipment for a fraction of the cost as long as it is in serviceable condition.

SOIL PREPARATION FOR PLANTING

Preparing the soil for planting is the most important step in crop production. Soil preparation consists of plowing, discing, harrowing, and dragging or firming the soil. It is important for planting small vegetable seed that the surface be fairly smooth, firm, and free of clods and trash. Thus the planter can place the seed at a uniform depth and cover it uniformly with soil. Large vegetable seeds, such as beans, corn, and peas, do not require as precise planting as small vegetable seeds.

Plowing should be to a depth of six to eight inches in most soils. It requires a great amount of power to plow to that depth unless a very small plow is used, and it is best handled with larger equipment. It is best to plow in the fall for early spring crops, and it is beneficial to grow a winter cover crop, such as rye, where crops will

Fig. 15. Winter cover crops of small grains, such as rye and wheat, can be planted in strips to protect young emerging seedlings from spring winds. The strips reduce wind speed and wind-blown soil that can damage young plants.

be planted in the late spring. Cover crops improve the soil and crop yields. They should be plowed under about a month in advance of spring planting to permit proper soil preparation. Do not plow when the soil is wet, particularly in heavy clay soils. When plowed too wet, heavy soils bake when they dry and are very difficult to get into good shape for planting. Likewise, if clay soils are too dry, large chunks will be plowed out that are difficult to break up before planting. A heavy soil is right for plowing if a ball that has been compacted by hand can be crumbled apart.

Soil should be disced and harrowed soon after spring plowing, but fall-plowed soil should be left rough over winter in areas where freezing occurs. A disc will cut and break clods and help level the soil. Heavier soils may require discing twice. A spike- or spring-tooth harrow should follow the disc harrow to level and smooth the soil further.

A drag may also be needed to crush remaining lumps of soil and finish the smoothing and firming operation after harrowing. The use of a drag may not be necessary in sandier soil.

Fertilizer must be applied to most soils to produce high crop yields. You must know your soil and the requirements of the crops that you are planning to grow before applying fertilizer. You need a soil analysis to learn your soil fertility level. State agricultural colleges do soil testing for a small fee. All you need to do is collect about a pint of soil from your field and take it to your county cooperative extension center, which is usually located in the county seat. The extension center personnel will get the soil analyzed and assist you in determining how much and what type of fertilizer to apply to your crops. It is a good practice to test your soil every two or three years, so that the correct amount of fertilizer is applied and none is wasted due to over-application.

The time and method of fertilizer application is important for vegetable crops. Attachments can be put on most vegetable planters to apply fertilizer near the seeded row, where the fertilizer will be used more efficiently by the crop.

It is best to apply fertilizer either before plowing or just before discing if a large amount is needed or your planter is not adequately equipped. All of the fertilizer needed may be applied before or at the time of planting except for nitrogen applied to crops during the growing season. For larger applications it is best to plow or disc in part of the fertilizer and then band the rest near the seed while planting.

Your soil-analysis report will also tell you the acidity of your soil. If it is too acidic, lime will help correct the condition. Soil can also be too basic for growing some crops. Sulfur or acid-forming fertilizer should be used if that is your problem. Soil acidity influences crop growth by regulating the availability of soil nutrients to the plants. Remember that you can obtain information concerning the need for lime or sulfur, the amount, and the best method of application in your soil from your county agricultural extension agent.

VEGETABLE TRANSPLANTS

Transplants are young plants grown, usually from seed, in greenhouses, hotbeds, cold frames, or sometimes open field beds. The success of a crop often depends upon the quality of the transplants used in its establishment. In many areas you may buy locally grown transplants at reasonable prices. You should consider whether you wish to master the techniques of growing transplants or are content to buy them from others.

Growing good-quality transplants is an art as well as a science. To begin, you must obtain good-quality seed of the varieties that are recommended for your area. Next in importance is the growing mix. Soil is used very little in modern plant production. Several good commercial plant-growing mixes are available from garden stores or greenhouse supply outlets. Some standard brands are Jiffy Mix, Redi Earth, and Pro-Mix. The growing media basically are peat moss and vermiculite. Other components are limestone and fertilizer. The mixes can be purchased ready to use, or the raw materials can be purchased and mixed. The following is a recipe for a multipurpose plant growing medium:

11 bushels of loose, shredded spagnum peat moss

11 bushels of horticultural vermiculite (sizes 2, 3, or 4)

2.5 to 5 pounds of pulverized dolomitic limestone

1.25 to 2.5 pounds of pulverized 20 percent superphosphate

1 pound of potassium nitrate or calcium nitrate

1 ounce of fritted trace elements (e.g., FTE 503) *or*
0.5 ounce Borax (11 percent boron) plus 1 ounce chelated iron (e.g., NaFe 138 or 330)

The seedlings should be fertilized with a complete fertilizer (4 to 6 ounces per gallon) when they are small and then at weekly intervals.

There have been several recent advances in growing transplants, involving the use of ready-made peat pellets and fiber blocks. These plant growing systems are usually more expensive than other growing media but very convenient. For example, the Jiffy 7 or Jiffy 9 peat pellet is popular for starting plants. When moistened, the small compressed pellet expands. The seed is then pressed into the peat. After the plant has grown, everything is transplanted. Similarly, fiber blocks are light, clean, and easy to use. After the roots have penetrated the block, the whole thing is planted. Manufacturers control the content of peat pellets and fiber blocks so that the plant growing containers are uniform, clean, and disease-free.

Follow strict sanitary practices in your greenhouse or transplant growing beds. Keep the area clean and do not allow soil to contaminate the growing mix. Use seed purchased from a reputable seed supplier because it is usually free of disease. Plant growing containers, such as greenhouse flats, should be new or thoroughly cleaned and disinfected before use. Greenhouse flats may be disinfected by soaking in a 0.26 percent sodium hypochlorite (chlorox) solution for five minutes. They must be allowed to dry completely before use. Use a commercial

Fig. 16. Mid- and late-season transplants can be produced relatively inexpensively in open field beds, but the open beds cannot be used for early-season transplants because they cannot be heated artificially. Field beds selected for production of transplants should have sandy soil and be well drained and free of troublesome pests. A southern exposure with some wind protection is also desirable.

soil-free plant growing mix, as discussed above. Since there is no soil in these mixes, many diseases are avoided. When watering young plants, keep the plants and the soil from remaining wet too long. To avoid overwatering, water plants in greenhouses or hotbeds only on sunny mornings. Ventilate plant growing structures to keep the relative humidity as low as possible and to provide air movement to keep the plants dry. Fungicide spray materials will also help prevent plants from rotting or damping off.

Peat pellets and fiber blocks are spaced in a greenhouse flat for easy handling. Then the seed is sown directly into the plant growing containers. If peat pellets or fiber blocks are not used, the small seedling plants are grown in seedling trays until they are about 1 to

1½ inches tall. They are then transplanted by hand to the plant growing containers.

To grow such small seedlings for plant growing containers, fill a greenhouse flat with your plant growing mix. Then make rows about one-half inch deep and two to three inches apart. Sow about ten seeds per inch in the rows. Cover the seed with fine vermiculite and water gently but thoroughly. Label each flat, showing the variety and the date of planting. Cover the flat with polyethylene film to prevent drying. Place it in an area with a temperature of 75° to 80° F. In a few days, when the seedlings begin to emerge, remove the polyethlene film cover and place the flat in full sun. Water the seedlings as needed, but do not overwater. When the seedlings are well developed (two or three weeks from sowing), transplant them to the growing containers. Most plants should be spaced about two inches apart. Water the seedlings thoroughly after transplanting.

Most transplants will do better in the field if they are hardened for seven to ten days before transplanting. "Hardening" consists of withholding water and lowering the growing temperature. The hardening treatment should not be too severe, but you do want the plants to slow their growth rate and toughen up to withstand the move to the field.

Vine crops (such as watermelon, muskmelon, squash, pumpkin, and cucumber) are never grown as seedlings and then transplanted to growing containers. They must be directly seeded into growing containers. The best containers are peat pots, fiber blocks, or any other containers that can be moved directly to the field and transplanted with the crop. Vine crops are more difficult to transplant than peppers or tomatoes, and their roots should not be disturbed during the transplant growing procedure. Transplanting is not used at all in the production of beans or sweet corn because those crops do not readily establish themselves after their root systems have been disturbed.

In table 22 is essential information on growing transplants, including the amounts of seed required.

PLANTING SEED AND TRANSPLANTS

A good seedbed must be prepared both for seed and for transplants, and both must be planted at the right time and the right depth and spaced properly in the row.

The best time for planting depends on the crop, the weather, the soil, and the time that production is desired. In a PYO operation it is desirable to have continuous crop production for as long as possible. This is attained by making successive plantings of crops such as beans and sweet corn at one-week intervals. The climate in your area will determine when planting can begin.

There are no definite rules concerning planting depth. Generally, large seed can be planted deeper than small seed. Sandy soils permit deeper planting than clay soils. Seed must be planted under the soil surface where soil is moist enough for seed germination. If irrigation water is available to prevent the soil from drying out, the risks associated with shallow seed planting are lessened. If seed is planted too deep, the seedlings may not be able to make it to the soil surface.

Plant rows as straight as possible. Straight rows are easier to cultivate and spray and generally look better. The space between rows should be uniform. Stakes or a line can be used to mark the first row. If your first row is straight, it is easier to keep the remainder straight. A row marker can be attached to your planter or tractor that will serve as a driving guide to keep rows straight

Table 22.
Information for Growing Your Own Vegetable Transplants

Vegetable	Average Number of Seeds per Ounce	Plants to be Expected per Ounce of Seed	Plants Needed to Transplant One Acre	Weeks Needed to Grow Transplants	Seed Germination Temperatures	Plant Growing Temperatures Day	Night	Square Inches of Space for Each Plant
Cabbage*‡	8,500	5,000	14,000	5–7	70–80	55–60	50–55	4
Cauliflower*‡	10,000	5,000	14,000	5–7	70–80	60–65	55–60	4
Broccoli*‡	9,000	5,000	14,000	5–7	70–80	60–65	55–60	4
Tomato*	11,000	4,000	3,000	5–7	70–80	65–70	60–65	4
Peppers*	4,500	1,500	10,000	6–8	75–85	65–70	60–65	4
Eggplant*	6,000	2,000	4,000	6–8	75–90	70–80	65–70	4
Cucumber†	1,000	500	3,000	3–4	70–95	70–75	60–65	9
Muskmelon†	1,000	500	3,000	3–4	75–95	70–75	65–70	9
Squash†	300	200	3,000	3–4	70–95	70–80	65–70	9
Watermelon†	300	200	1,500	3–4	70–95	70–80	65–70	9

*Grow in greenhouse flats. Individual plant containers are not needed.
†Use a plant-growing container, peat pots, or fiber blocks. Seed directly into plant-growing container.
‡Can be grown in open field beds for fall transplanting.

and evenly spaced. Standardized row spacing for your crops eliminates the need to keep adjusting your planter and cultivation equipment. Most vegetable crops can be grown at 36-inch row spacings. Most small tractors can straddle two 36-inch rows for field operations. Tomatoes and vine crops need wider row spacing, and twice 36 inches, or 72 inches, is a convenient row spacing for growing them.

There are many good machines to plant vegetable seed. The Planet Jr. is one commonly used on small vegetable farms. It is versatile and can be used to plant the seed of any vegetable crop. A Planet Jr. is also much less expensive than the more complex machines used by large vegetable growers. The function of a planter is to open a furrow, drop the seed, and cover and firm the soil over the seed. Planters are adjustable for different planting depths and different rates of seed drop.

How thick to plant seed depends upon several factors, including how good your seed is (the percentage that will germinate), which is printed on the seed container; the soil conditions; the time of planting; and the expected losses to insects and diseases. Seed with a low germination percentage must be planted more thickly. Seed should also be planted more thickly if soil or weather conditions are not optimum. If unusual insect or disease attacks are anticipated, more seed should be planted. The final stand that emerges from the seed planting should be close to the desired stand. If too many seeds produce plants, some may have to be thinned to ensure a good quality crop and a more uniform stand of plants. The weaker, unwanted plants should be removed first, and the thinning should be done when the plants are small to avoid overcrowding and injury to the plants that are left.

Some vegetable crops, such as tomato, cabbage, and

pepper, should always be transplanted. To ensure success, you should acquire good transplants, have the soil in good condition, and do the transplanting job correctly. The plants can be set in the field by hand, but a transplanting machine will save much time and labor on larger acreages.

If the plants are set by hand, the furrows should be opened with a cultivator shank or other tool at the proper row spacing, which also marks the row. At each planting site in the furrow a hole is made by hand or with a trowel before the plant is placed in the hole. The roots are then covered with soil, and the soil firmed around the roots. The remainder of the furrow can be filled by using a tractor cultivator. A pint of water containing a small amount of starter fertilizer should be poured around each plant.

A transplanter is pulled by a small tractor. It opens the furrow, applies the water, sets the plant, and firms the soil in one operation. Usually two or three people are needed to operate a transplanter, one to drive the tractor and one or two to feed plants into the machine.

Set plants slightly deeper than they were growing before transplanting. Tall spindly plants should be set deeper still to protect them from wind damage. Strawberry transplants must not be set too deep, however, because the plants are likely to die if their crowns are covered with soil.

In tables 23 and 24 is information concerning times when different crops should be planted and the quantities of seed needed to seed one acre directly in the field.

Table 23.
Vegetable Seed Needed to
Seed One Acre Directly in the Field

Vegetable	Pounds of Seed per Acre*	Vegetable	Pounds of Seed per Acre*
Snap bean	60–100	Onion	3–4
Beet	8–10	Pea	60–100
Broccoli	1.5–2	Pepper	2–4
Cabbage	1.5–2	Pumpkin	2–3
Carrot	2.5–3	Radish	9–10
Cauliflower	1.5–2	Southern pea	30–40
Cucumber	1.5–2	Spinach	10–12
Eggplant	1.5–2	Summer squash	4–5
Lettuce	1–2	Sweet corn	10–12
Muskmelon	1.5–2	Tomato	1–2
Okra	5–7	Watermelon	1–2

*Seed needed for direct field seeding will vary with planting conditions and row spacings. When planting conditions are favorable, the lower seeding rates can be used.

Table 24.

A Planting Guide to Crops According to Their Tolerance or Need for Cool or Warm Weather

Cold-Weather Plants for Early-Spring Planting		Cool- and Warm-Weather Plants for Late-Spring or Early-Summer Planting			Planted in Late Summer (1–2 Months Before First Fall Frost)
Planted 4–6 Weeks Before Frost-Free Date	Planted 2–4 Weeks Before Frost-Free Date	Planted on Frost-Free Date	Planted 1 week or More After Frost-Free Date	Planted in Warm Weather	
Asparagus	Beet	Squash	Snap bean	Bean	Beet
Blackberry	Carrot	Sweet corn	Eggplant	Chard	Broccoli
Blueberry	Chard	Tomato	Pepper	Southern pea	Cabbage
Broccoli	Mustard		Sweet potato	Squash	Cauliflower
Cabbage	Parsnip		Cucumber	Sweet corn	Collard
Cauliflower	Radish		Southern pea	Okra	Kale
Lettuce			Squash		Lettuce
Onion			Sweet corn		Mustard
Peas			Pumpkin		Spinach
Potato			Muskmelon		Turnip
Spinach			Watermelon		
Strawberry					
Turnip					

PEST CONTROL

Good pest control is necessary to provide an abundant supply of high-quality produce throughout the growing season. Pests include diseases, weeds, insects, nematodes, and other animals.

Weeds are a problem to growers everywhere. PYO customers will not search through weeds to find something to pick. If weeds take over a planting, you must either clean the weeds out by cultivation or hand labor or destroy both the weeds and the crop by mowing, discing, or plowing.

Insects and diseases can blemish or completely destroy a crop if not controlled. It is a joy to pick high-quality, blemish-free produce: make your customers happy so that they will come back for more by effective pest control.

Nematodes are microscopic worms that damage the roots of plants. If your plants have a nematode problem, you will only see the results in poor growth and low yields.

Animals can be a problem in some crops. Birds love berries and shiny, ripe tomatoes, for example, and you are going to incur some losses to birds no matter what you do. You may need to use a noise maker to scare them away if losses are too great. Snakes can cause complete disaster in a PYO field because many customers are scared silly by a snake. The sound of someone screaming "Snake!" will clear a field of customers in seconds—and stampedes can damage crops. Good weed control is the key to keeping snakes out of fields. If there are not a lot of weeds and trash for them to hide under, they will stay away.

It is very difficult to operate a PYO without the use of some pesticides, but commercial chemicals are not

Fig. 17. This weedy snap-bean field will not attract PYO customers. Snakes like weedy areas, and nothing will clear a field of customers faster than a frantic scream of "SNAKE!" Weeds also attract insects and diseases that attack crop plants.

the only means that you should use in pest control. In fact, they should be your last resort. Farming practices such as crop rotation and good field sanitation are very beneficial. Plow or disc your crop residues immediately after harvest, whether or not they are infested with insects or diseases. Use resistant varieties whenever possible. Stay out of your fields when plant foliage is wet, and manage irrigation properly so that plants do not remain wet all night. You will encourage beneficial insects by minimizing sprays. You can purchase lady bugs through the mail for release in your crops to help control aphids.

Weeds cause many problems. They reduce crop yields and increase insect and disease problems that cost you money. Crop plants generally produce lower-quality products when they compete with weeds because the weeds use valuable soil nutrients, light, and water that the crop plants need. A special consideration for PYO farmers is that some customers have allergies, and weed pollen is the principal cause of hay fever.

Controlling weeds is one of the major costs in crop production. There are several basic principles to consider to reduce weed-control problems and costs. You should never attempt to establish vegetable or small fruit crops in a field that is badly infested with perennial weeds, such as bermudagrass, Johnsongrass, quackgrass, nutsedges, and thistles. Herbicides or tillage should be used to control the weeds at least one year in advance of planting a crop.

Weed control usually requires a combination of several management techniques. If possible, grow different crops on the same area each year. Growing the same crop year after year, using the same weed control techniques, will encourage certain problem weeds. The rotation of crops and the use of herbicides and tillage methods will help prevent problem weeds from developing. Whenever you see a severe weed problem, it should be eradicated immediately even if it is only in a small area. Weeds should be prevented from producing seed whenever possible. One plant can produce thousands of seeds, and the seeds will usually live in the soil for years.

Although chemicals control weeds effectively, certain risks are involved in their use. Crop injury is one risk. No crop is completely resistant to herbicide damage, though each is able to tolerate a certain dosage range. Selectivity, or the ability of a herbicidal chemical to kill weeds without harming crop plants, may be partially lost under certain weather conditions. Careless application can also injure your plants or neighboring crops. Herbicidal injury can range from complete destruction

114

of plants to slight stunting or discoloration, which are usually accompanied by lower crop yields.

Good weed control is mandatory in PYO operations. Use chemicals wisely, have your cultivator adjusted properly, and make sure your hoe is sharp. All of these tools will be needed.

Like weeds, insect and disease problems are nearly always present. Unlike weeds, which are easily visible in the field, insects and diseases are often difficult to see. Crop plants must be looked at closely on a daily basis to discover insect and disease problems before they become severe. If you do not check your crops frequently, it may be too late when you do finally discover a serious problem.

Insecticides to control insects, fungicides and bactericides to prevent or stop diseases, and nematicides to control nematodes—all may be necessary. Remember, however, that chemicals should be your last resort. Many other crop production practices will help prevent problems from developing to the point where they require the use of pesticides.

A good, well-planned crop rotation system is very helpful in preventing insect and disease problems. Do not plant your tomatoes or other crops back into the exact same plot of land each year. To do so will only encourage a buildup of insect and disease problems on the crop. Choose resistant varieties where possible. For example, tomato varieties are available that are resistant to many diseases and even to nematodes. If you take advantage of the disease-resistant varieties that are available for all the crops that you grow, you will reduce your disease problems.

Another practice that you can use to reduce disease and insect problems is good field sanitation. Plow under

Fig. 18. Diseased plants are unsightly and produce low yields of poor-quality produce. Foliage diseases have drastically reduced the yield in this tomato planting. If foliage disease pressures are as heavy as this, even resistant varieties may need to be protected by timely applications of approved pesticides to avoid complete crop loss.

the trash and debris left by the previous crop. This old plant material is good in the soil, but if it is left on the surface, it just harbors insects and diseases that will be waiting to attack your next crop. Many diseases need moisture or high humidity to attack plants. Irrigation by sprinkler and other methods increases humidity and wets plant leaves, providing an ideal situation for diseases to attack the plants. Some of these problems can be avoided if irrigation is managed properly. Irrigate during the day so that plants will dry quickly. The shorter the time when plant leaves are wet, the less the likelihood that diseases will get started. Irrigating in the late evening or at night can be disasterous because the plants will probably remain wet all night.

Many insect- and disease-control chemicals are available for use on crops, and some will undoubtedly be

Fig. 19. Rotation of vine crops is very important to avoid *Fusarium* wilt disease. The muskmelon plants shown here are wilting and will soon die without producing a crop. Fields in which diseased plants are observed should be planted to a nonrelated crop in the following season.

necessary in your operation. Advanced planning is needed before a chemical is used, particularly where several different fruit and vegetable crops are being grown. You cannot be spraying chemicals for insect and disease control when customers are around the field. It must be done on days you are closed or in the early morning or late evening, and you must carefully take into consideration the waiting period before customers can safely enter the field after pesticide application. There will also likely be a waiting period after application of a chemical to a crop before the crop can be safely harvested. The waiting period may be less than one day or as many as fifteen or more days.

Insect sprays kill not only the bugs that you are after but also many beneficial insects. Honeybees are very beneficial because they pollinate several different fruit and vegetable crops. Be sure to keep bees in mind when you are using insecticides. Ladybugs and assassin bugs kill large quantities of insects, and it is self-defeating for you to kill them by indiscriminate use of insect sprays.

With those factors in mind, extreme care must be taken in the use of pesticide chemicals. Every commercially marketed chemical-pesticide container has a label, and all the safety information concerning the chemical is on the label. Read the label completely before using the chemical on a crop.

Before pesticides are marketed, a tremendous amount of research is conducted to determine how they may safely be used. Do not fail to use this information to protect the environment, yourself, your crops, and your customers.

MULCHING VEGETABLES

A mulch is any substance spread over the soil surface to protect plant roots from cold, heat, or drought or to keep the plant fruit clean. Increased yields, earlier harvests, and fewer rotted fruits are obtained by mulching. Mulching also reduces weed growth and keeps fruit and vegetables cleaner. Mulch can be of particular value in growing tomatoes in the ground without staking and in growing muskmelons, cucumbers, and eggplants. Mulching materials can be plastic, paper, straw, sawdust, or the residues of other plants, but the mulch most used by vegetable growers is plastic film. Other materials are just too expensive, particularly if they must be hauled very far.

Plastic mulch is made of very thin polyethylene film.

Fig. 20. Warm-season crops, such as tomatoes, peppers, melons, and cucumbers, produce earlier when they are grown on plastic mulch. Black plastic film is the most common mulch used in vegetable production. It is relatively inexpensive, and the laying of the film is completely mechanized. Transplanters and seeders are available that will plant through plastic mulch.

The rolls of film come in widths three to seven feet wide and are usually 1,000 or 2,000 feet long. The three-foot width is fine for most of your warm-season crops. Plastic mulch is relatively inexpensive (costing about $200 per acre) and can easily be placed in the field by a mulch-laying machine. It comes in various colors, but black is most commonly used. The big headache in using plastic mulch is that it must be removed from the fields at the end of the growing season. If it is not removed, it will remain in the soil and get tangled in cultivating and planting equipment in future years.

The soil is prepared for use with a plastic mulch in

the normal manner for transplanting. The plastic mulch is laid over each row after soil preparation. The soil warms under the mulch, causing transplanted crops to grow faster in the spring. A hole is punched into the mulch to make a planting hole in the soil for the transplant. Rain water and irrigation water get under the plastic through the transplanting holes and by moving in from the edge of the mulch.

CROPPING SYSTEMS

For the most efficient use of your land, you must plan your crop production one or two years ahead. After you have been in operation for a season or two, you will have a good idea about which crops your customers will want and how much you should grow. You will also know which crops you can or cannot grow and which crops are making you the most money.

With this information you should plan where each crop will be planted on your farm, following a crop rotation system. Crop rotation is a systematic plan for growing different crops on the same land each year. A rotation plan should avoid having the same crop planted in the same area more than once every three or four years. A good crop rotation system will help keep insect, disease, and weed problems to a minimum and will also make your soil perform better.

More efficient and profitable use of your land can also be realized by succession cropping, sometimes called double-cropping, whereby more than one crop is grown during a single season on the same land. Succession cropping is easier in areas with longer growing seasons, but can be done in areas with shorter growing seasons. Since many vegetable crops occupy the land for relatively short periods of time, they can be followed by a second or even a third crop. An early spring crop, such as cabbage, lettuce, or beets, might be followed by beans, and the beans might be followed by a fall crop of turnips. Numerous combinations can be worked out to fit your needs. You will want to maintain some flexibility in your schedule because the best-made plans must sometimes be changed if weather conditions are unusual. An example of successful succession cropping is the one-acre surprise described at the close of Chapter 4.

You should also know about intercropping. In this cropping system two or more crops are grown on the same land at the same time. Intercropping is more difficult to manage than succession cropping, but it does make very efficient use of the land. The rows of perennial crops, such as small fruits and asparagus, usually are wide, and space is available between the rows to grow another crop. When considering intercropping, you must plan on the increased labor that will be required because your cultivator and other equipment may not fit the rows. More water and fertilizer are needed because two crops are grown instead of one. Controlling pests may also be more of a problem if the same chemicals cannot be used on the adjoining crops.

Generally, a small, fast-growing crop is interplanted with a crop that has a longer season or with a young perennial crop. One possibility would be to intercrop early spring cabbage and tomatoes. The cabbage would be transplanted into the field in three-foot rows early in the season. It would be nearly ready to harvest at the time when the tomatoes should be transplanted. The tomatoes would be transplanted into every other one of the cabbage rows, which would be spaced six feet apart. Before the tomatoes had grown very large, most of the cabbage would be harvested and would not interfere with the growth of the tomato plants. Rows of new asparagus and small fruit plantings are also six feet or more apart, offering space for intercropping, but when an asparagus or small-fruit planting has developed and is producing, the competition from the intercrop is likely to reduce yields. Generally, intercropping should only be attempted during the first year that small fruits are planted.

In table 25 is information concerning the number of days after planting before the harvest of eleven PYO vegetables. With this information you can plan your production and determine what crops may work for you in successive-planting or intercropping systems.

Table 25.

Days Required from Planting to Harvesting Vegetables

Vegetable	Days to Harvest*	Vegetable	Days to Harvest
Snap bean	45–65	Onion	70–100
Beet	50–75	Pea	60–75
Broccoli†	55–100	Pepper†	60–80
Cabbage†	60–100	Pumpkin	100–120
Carrot	60–85	Radish	22–40
Cauliflower†	55–120	Southern pea	60–80
Cucumber	50–60	Spinach	40–50
Eggplant†	70–80	Summer squash	50–65
Lettuce	40–80	Sweet corn	65–85
Muskmelon	80–90	Tomato†	65–90
Okra	50–60	Watermelon	85–100

*A range of days to harvest is given here. The lesser number of days is for early-maturing varieties in good growing weather. Days to harvest will be more when the crop is maturing at cool fall temperatures.

†For transplanted crops the days-to-harvest figure does not include the time required to grow the transplants.

8

PYO Crop Production Tips

STRAWBERRIES

Strawberries are a desirable dessert fruit and very nutritious. One cup of fresh strawberries gives more than the recommended daily allowance of vitamin C. They are also versatile fruits used fresh and frozen in desserts or preserved in jellies or jams. For those reasons the strawberry is probably the most widely grown crop in PYOs.

Although strawberries are a highly rewarding crop to sell in a PYO, they do require more attention than other crops. Following good horticultural practices will produce yields of 5,000 to 10,000 quarts of strawberries per acre. Well-cared-for plantings will profitably produce strawberries for at least two to three years.

Strawberries must have good soil drainage. They cannot tolerate standing water. Soil that is too wet reduces plant growth and encourages rotting of the roots. Because strawberries bloom early in the spring, do not plant them in a low-lying area that is susceptible to frost. A late spring frost would destroy the early flowers and ruin your crop. Nor should you plant strawberries on a steep slope where heavy rains will erode or wash out the plants. If you pick an area recently planted to potatoes or tomatoes or growing in grass, you will have too many disease and pest problems.

Several varieties of strawberries should be planted to extend the harvest and profit season. Late spring-bearing strawberries produce one crop a year, while everbearing strawberries produce one crop during the normal season and another crop in the fall. It is important to select varieties suited to your climate and soil conditions. If you grow varieties that are well-known in your area you may be able to get higher prices. Many buyers prefer picking the berries of just one well-known variety that are uniform in color and size.

Buy from a reputable nursery to ensure that you will get quality plants. Order the plants as early as possible and indicate the date when you will need them. The word "registered" on a plant label indicates that the plant has been grown under state supervision and is the best that can be purchased. Registered plants are virus free and may yield 50 percent more fruit than ordinary plants. There are many registered virus-free varieties available. Plants may also be "certified", which means they are free of most insects and bacterial diseases but not viruses. Of some varieties the certified plants are the best available.

Dormant strawberry plants are often used in spring planting. Dormant plants that have been properly stored are as good as freshly dug plants.

When you obtain your strawberry plants, check them and moisten the roots if necessary. Plants may be stored for several weeks if they are kept between 32° to 36°F. Leave them in the plastic bags that they came in.

Better results are obtained by preparing your soil the year before planting to control weeds and adjust the fertility level. This needs to be done in late summer or early fall. Manure may be added at this time if it is available. Strawberry plants should be set out in the spring as soon as the soil can be prepared, but not when the soil is too wet.

As early as possible in the spring work the soil until it is nearly in seedbed condition. Fertilizer may be applied and worked in at this time too, but fertilizer is not recommended in the spring of harvest years because it causes soft berries and reduces yields. Chemicals to control soil insects may be applied and worked in with the fertilizer.

Keep in mind how you plan to train the strawberry runner plants when you determine the distances between rows and between plants. Your customers will need plenty of space, but your rows should not be so wide that they cannot pick from both sides of the row easily. Most varieties are set every twenty-four inches in rows at least forty-eight inches apart. An acre will require 5,000 to 6,000 plants with that row spacing.

If a transplanter is not available, use a shovel or spade to make holes about four inches deep for the plants. The plants will have been kept moist before being set out. Carry the plants in a container, such as a bucket or basket, to prevent them from drying out. Remove all but two or three of the hardiest leaves. Spread the roots on the soil and cover them until just the crown of the plant is above the soil surface. If the roots are exposed or the crown is covered with soil, you may lose your plant. Firm the soil around the plants with your foot before giving each plant at least one cup of water. It is most efficient to have one person to dig the holes, another to set the plants, and a third to apply the water. Make sure your plants receive plenty of water and do not dry out after planting. Vigorous growth is necessary to produce large quantities of quality fruit.

126

During the first season pinch off the strawberry blossoms as they bloom, or they will reduce plant growth and the next year's crop. Keep the plants cultivated once a week during the first season to control weeds, but do not cultivate deeper than one inch near the plants. Mulching in the spring of the fruiting year helps control weeds and allows runner plants to root more easily. Keep the rows eighteen to twenty-four inches wide.

Runner plants emerge where leaves join the main stem. They form new plants and take root several inches from the original plant. Varieties differ greatly in the number of runner plants that they produce from each original transplant.

Winter mulching is recommended to protect strawberry plants from severe cold in northern areas. Straw is the best mulching material because it is clean and lightweight. Place the straw near the plants early in the fall so that any grain or weed seed will germinate before the mulch is placed on the strawberries. After several sharp freezes spread the mulch four inches deep over the plants. Crown damage may occur if mulching is done before growth stops in the fall. When new leaves develop in the spring, take the mulch off with a rake or pitchfork and place it between the rows. This controls weed growth, retains soil moisture, and keeps the berries clean. If frost is predicted, the mulch can be spread back over the plant rows to protect the flowers.

Strawberries should be harvested when they are fully colored. The first harvest is usually about thirty days after the first flowers bloom. Berries are usually harvested every other day, but do try to avoid picking when plants are wet. The harvest season of a variety normally extends over six or seven pickings.

Locations that have sunny days and cool nights produce strawberries with more flavor. Flavor is not as good when the days are cloudy and humid and nights are warm. Temperature also has a great effect on the firmness of the berries. Strawberries grown in cool weather are firmer than those grown in warm, humid weather. Temperature also effects the ripening season. Higher temperatures speed ripening and shorten the time between blossoming and ripening. Cool weather may delay ripening several days.

It is not profitable to try to get more than two or three years of production from a strawberry planting. The size of the berries and the yields are progressively smaller each year. Therefore make new plantings each year and destroy your oldest fields. Customers are much happier when picking large berries.

Immediately after the harvest season renovate each planting by mowing off the rows as close to the ground as possible. Then narrow your rows to about ten inches and thin the space between the plants in the row to three or four inches. When narrowing the rows, work your mulch into the soil and then apply fertilizer. Treat the renewed plants the same way you would a first-year planting.

You can obtain new plants from an established bed by transplanting the more vigorous young plants growing along the edge of the rows in early spring. Remember that diseases can build in strawberry beds. Obviously you should not use your own plants if they are diseased. Go instead to your nurseryman and purchase good disease-free plants. Most disease and pest problems can be easily managed by obtaining good plants and following good cultural practices.

Everbearing strawberry varieties will produce fall berries, but they require a lot of labor and do not produce well in the matted rows discussed above. The benefits do not justify the extra labor. They are not recommended for a PYO unless the planting is of a limited size.

BLACKBERRIES

A PYO may be the only source of fresh blackberries for many consumers. A blackberry plot may yield 6,000 quarts of berries per acre if managed properly. Blackberry plants are easy to grow, require little work, and will begin producing berries in the second year after planting. The plants are fairly hardy, withstanding winter temperatures as low as -15° F, but they grow best in temperate areas.

There are two types of blackberries, erect and trailing. The difference is in their canes, or branches, and in their fruiting characteristics. Erect blackberries have self-supporting canes. The canes of trailing blackberries must be supported by some means, such as poles or trellises. The fruit clusters of trailing blackberries are more open, ripen earlier, and usually are larger and sweeter.

You must have a good water supply in order to grow blackberries. They require a lot of water during the growing and ripening period. They are best watered either by soaker hoses or drip irrigation. Thus the foliage is not wet for long periods of time, making them susceptible to diseases. Blackberries will do well in most types of soil if there is good water drainage.

Blackberries can be protected from the weather by planting them on hillsides to avoid winter damage and late frosts. Surrounding trees or hills will protect them from wind.

It is advisable to seed a green-manure crop, such as rye or vetch, and plow it under to condition the soil for planting. Plow about nine inches deep as early as the soil is workable in the spring. Disc just before setting the plants.

If your plants are dry when you receive them from your nursery supplier, soak the roots for several hours before planting. If they cannot be planted soon, protect them by heeling in. This is done by digging a hole large enough for the roots. The plants are spread out in the hole with the roots down. Then the roots are covered with moist soil. While they are being planted, their roots should be protected by plastic bags to prevent drying. Cut the plant tops down to six inches so that they may be used as handles as you set the plants out. The plants should be set the same depth as they were in the nursery. With your heel pack the soil firmly around each plant. For cultivation erect blackberries need to be planted about five feet apart in rows that are eight feet apart or wider. Space trailing blackberries about eight feet apart in rows ten feet apart.

Although erect blackberry plants are self-supporting, their canes may be broken during picking. Trellises will pay for themselves because they reduce damage considerably. A simple trellis can be made by stretching wire between posts about fifteen to twenty feet apart in the row. Trailing blackberries need two wires for support. One wire should be placed three feet from the ground, and the other five feet. One wire about three feet high is sufficient for erect blackberries. Use soft twine to tie the canes to the wire. Train your trailing plants to grow horizontally along the wires.

Blackberry plants will need to be cultivated frequently during the first half of the season to control weeds. Begin cultivating early in the spring and continue on a weekly basis until the end of August. Cultivate only one to two inches deep near the rows or you may harm some of the roots. Fertilizer may be applied once a year at blossoming time to maximize yields.

Canes that grow from crowns live two years. The first year they send out side branches. The second year small branches grow from buds on the older branches. The cane dies after the fruit is produced. Prune the branches back to twelve inches in the spring before new growth begins. These methods will produce larger blackberries of higher quality.

Erect blackberries produce root suckers as well as new canes. All of these suckers need to be cut off to keep the plant growth from becoming unmanageable. When your erect blackberry plants reach thirty-six inches in height, cut the tips off the canes to make them branch. They then will provide more support for the berries.

Customers may pick blackberries as soon as they become sweet. Firm and ripe berries fetch the best prices. The berries will keep better if they are picked early in the day. They need to be picked about every other day. Blackberry plants usually average over a quart of berries per plant.

Upon completion of harvest, cut down all the old canes. The new canes may be thinned to four canes for erect plants and eight canes for trailing plants.

Propagating blackberry plants is a simple procedure, and most growers produce their own plants this way. If the tips of the canes of trailing blackberries are buried three inches deep in late summer, they will root and form new plants. Erect blackberries are propagated in early spring by digging pieces of root from established plants and planting them three inches deep. Cut the roots in three-inch lengths before planting.

Blackberries may be protected from drying winds during the winter by a winter cover crop, which will also prevent soil erosion. The crop is put in with the last cultivation by drilling spring oats or rye in the middle of the rows between the plants. The cover crop is destroyed by discing in the spring.

Canes can be protected from severe cold winters by bending them to the ground and covering the tips with soil before the ground freezes in the fall. Canes may also be covered with mulch. Uncover them before the buds start growing in warm weather.

You may have a few birds wanting to share the berries with your customers. Several possible solutions come to mind, such as bird netting or a noise maker, but you might also just have a few extra plants and expect a small loss to the birds. Most other pest problems can be prevented by following good cultural practices.

BLUEBERRIES

Blueberries cannot be grown everywhere in the United States and are sometimes regarded as difficult to grow. Yet observing just a few established practices will produce good plant growth and high yields. Because comparatively few people have the opportunity to taste the flavor of fresh blueberries right off the bush, they may be one of the most profitable crops in your PYO.

There are several different types of blueberry plants. Some are low and creeping, and others are upright and tall. They vary from two feet to eight feet in height. Many wild blueberries are harvested commercially for processing and home use. Cultivated blueberries produce berries that are several times larger than wild berries, and for that reason the cultivated varieties are preferred for fresh markets. The two main types are the highbush and rabbiteye varieties.

Highbush Blueberries. The native habitat of the highbush blueberry extends from southeastern North Carolina to southern Maine and west to southern Michigan. Its southern growing limit is 300 miles north of the Gulf of Mexico from Georgia to Louisiana. The highbush plant averages eight to ten feet in height and grows best where the water table is high and there is good surface drainage. The berries have better flavor in northern regions where the days are longer and the nights are cool. Highbush varieties are normally planted where more than five months of the year are without frost. Thirty-three to fifty days of the year should be below 45° F in winter but not below -20° F. In extremely cold areas deep snow may provide protection so that highbush blueberries can be grown. Highbush varieties cannot tolerate the very short winters in the extreme south.

Highbush blueberries tend to grow better in acid soils that are sandy with some loam. The soil must be very acid for the best growth, but add limestone if the soil is too acidic. If soil needs more acidity, fertilize with finely ground sulfur or ammonium sulfate. Use sulfur to increase the acidity of soil that is loamy and ammonium sulfate to furnish nitrogen in sandy soils. If you have a heavier soil, you need to add organic matter to make it more porous and improve drainage. Highbushes also require full sunlight to develop good flower buds. Good air circulation in the spring protects the flower buds from early morning frosts.

Plant highbush varieties about five feet apart in rows that are ten feet apart. Highbushes can set fruit with their own pollen, but two or three different varieties should be planted together to get cross-pollination by bees. Cross-pollination increases the size and numbers of the berries and shortens the ripening period. If native bees are inadequate, supplement them with your own bee colonies.

Rabbiteye Blueberries. Rabbiteye blueberries are native to southern Georgia, southern Alabama, and northern Florida. They grow mainly in areas that have a short winter. They are more heat and drought-resistant and require only sixteen to twenty-five winter days below

133

45° F. They are not so sensitive to soil type as the highbushes and can grow in upland areas. The rabbiteye bushes should be planted about six feet apart in rows twelve feet apart. More than one rabbiteye variety must be planted together to obtain pollination. The fruit of rabbiteye varieties usually ripens later and over a longer period of time than that of highbush blueberries.

Purchase two-year-old rabbiteye plants that are certified or state-inspected. Prune them to three or four shoots evenly spaced around the crown. Prune back the remaining shoots to remove the fruit buds. Set each plant in a hole about twelve inches deep. Press the soil firmly around the plant with your feet and water it thoroughly. Add mulch several inches deep for eighteen inches around the plant. This is to control weeds, keep the soil cooler in the summer, retain soil moisture, and control erosion. Dormant plants normally can be planted in the spring as soon as the soil can be worked. Blueberries can also be planted in the fall if your area has a mild winter.

Fertile soils only need a little fertilizer to maintain vigorous rabbiteye growth. More fertilizer is needed on poor soils, and it should be applied after the buds start to open. If the soil is not acid enough in an established planting, apply 100 pounds of ammonium sulfate per acre and repeat that application one or two more times at six-week intervals. Apply fertilizer evenly around each plant. Remember, however, blueberries are very sensitive to excess fertilization. The general rule is one ounce of fertilizer per year for mature plants, but extra fertilizer is needed to decompose mulch. During the first two years divide the fertilizer into several applications. After the third year apply the fertilizer only when the flower buds are beginning to open.

The root systems of blueberries are shallow. Normally

they are in the upper eight to ten inches of soil. The bushes should be cultivated very shallow so that roots will not be damaged. Such light cultivation improves soil aeration and controls weeds. If the area is mulched, pull the weeds by hand or hoe gently.

Blueberries will not resist drought. Therefore irrigation must be provided. To obtain maximum growth, the bushes need about one inch of water a week. Watering is most critical just after setting the plants and during the first two growing seasons. Good drainage is also essential, however. If the plants are standing in water, root damage will occur, and the plants may die.

Mature blueberry bushes should be lightly pruned each year during their dormant period to lengthen bush life and produce vigorous new growth. Very little pruning is needed until the end of the third year. Cut out low branches next to the ground, leaving just the erect branches. Cut out the older and weaker stems in the middle of the bush. Cut back smaller branches leaving the stronger shoots to branch. Pruning is normally done any time after the plants have gone dormant in late fall or winter. Berry size will be increased by pruning because the largest fruit is on the more vigorous wood. Pruning can also regulate the ripening time; the heavier the pruning, the earlier the berries will ripen. Thus pruning reduces the total crop for the year but increases berry size, hastens ripening time, and produces strong new growth for the next year's crop.

The harvesting period usually lasts six to seven weeks, during which three to seven pickings are made at weekly intervals. Few blueberries are produced during the first two years after planting. The maximum yields are attained in six to ten years. Between the second and the sixth year the yields may increase from 5,000 to 6,000 pints per acre. Mature blueberry plants produce fourteen to sixteen pints of berries per plant with moderate pruning. Well-managed blueberry plants will have between ten and fifteen prime bearing years.

TOMATOES

Next to the potato, the tomato is the most important vegetable raised and used in the United States. Tomatoes have many uses and have a unique flavor enjoyed by most people. You should certainly plan to plant tomatoes in your PYO even if you start the operation with only two or three different crops. Tomatoes can be grown with minimal effort and require little space for a large amount of production. Each tomato plant should yield from eight to twenty pounds or more of fruit. Home canning of tomatoes is popular, and many customers will purchase large quantities.

Tomatoes are a warm- and long-season crop. The plants will be killed by a hard freeze in the spring or fall. By using transplants to establish the tomato crop, it is possible to grow tomatoes in all areas of the United States. Almost any kind of soil will produce tomatoes. If your growing season between spring and fall frosts is short, you should grow tomatoes on a sandy soil for faster production. It is essential that the soil for growing tomatoes be well drained. The plants will die if the soil remains waterlogged for more than a few days.

The time and method of fertilizer application is important in growing tomatoes. Relatively large quantities of fertilizer should be plowed or disced in when preparing the soil. All of the phosphorus and potassium needed can be plowed in, but only about half of the nitrogen. The remainder of the nitrogen should be placed beside the tomato rows when the first fruits are set on the plant. Applying all the nitrogen before planting may cause a large amount of foliage growth, but it may also delay the harvest and reduce yields. The water used at transplanting should contain some fertilizer to get the plants off to a fast start.

Since tomatoes are easily damaged by frost, the plants should not be set in the field until the danger of freezing is past. If you want to try for very early production, use plastic mulch or purchase hot caps to place over each plant for protection against frost. Because hot caps are expensive and require labor in their use, they may not be worth the effort unless a higher price can be obtained for the early tomatoes.

Tomato plants are commonly set in the field in rows 6 feet apart with the plants from 2 to 2½ feet apart in the row. Early varieties and varieties that produce small bushes can be set at the closer spacing. Usually about 3,000 to 4,000 plants are needed per acre.

Tomatoes are flexible in that they can be grown either on the ground with no training and pruning or with elaborate staking or trellis systems. The labor required to train and prune tomatoes is very high, but much higher production can usually be attained in areas with high humidity if the plants are staked up off the soil. Foliage diseases can be a severe problem when a tomato plant lies on the soil. Fruit rots can also cause many of the tomatoes to be worthless if they touch the soil. Higher-quality fruit is generally obtained by training and pruning the tomato plants.

Good chemicals are available to control weeds in transplanted tomatoes. Shallow cultivation will be helpful in destroying any weeds present and to break up soil crust. Cultivation is limited where the vines of untrained plants cover the soil. Do not cultivate when the foliage is wet because that may spread diseases around your field.

Tomatoes will respond to irrigation when rainfall is

short and the soil becomes dry. Since the roots of the plants are deep in the soil, two or three acre inches of water should be applied at each irrigation. You should make sure that the soil is well supplied with water before the tomatoes begin to ripen, because irrigation water can increase the number of tomatoes with cracks if it is applied while the fruit is ripening. You should avoid irrigation after you have begun your harvest if possible.

Pruning

It is common to prune and train tomatoes in the humid regions of the southeastern United States. The plant is pruned to a single stem, which is tied to a stake. All the sucker shoots that grow where the leaves are attached to the main stem are pinched or broken off when the sucker shoots are small. This should be done three or four times at ten-day intervals. After each pruning, the plants are tied to the stake with twine. Tie the twine first to the stake and then loop it around the stem under a leaf. By this method you avoid tying the string tight around the stem, which may restrict growth or cut the stem.

A variation of single-stem pruning allows the first sucker shoot formed to remain on the plant and form a second main stem. This is called pruning to two stems. All subsequent sucker shoots that form on the two main stems are removed. Both of the main stems are tied to the same stake for support. Sometimes even three stems are allowed to develop to provide higher yields.

Total yield per plant will usually be less if tomatoes are pruned, but higher percentage of the tomatoes will be marketable. A closer spacing of plants in the row will make up for some of the lost yield due to pruning. The big disadvantage is the greatly increased cost of production because of the stakes and twine and large amount of labor required.

Customers will enjoy picking tomatoes that are staked, and the tomatoes will be cleaner. On the other hand, blossom-end rot and fruit cracking may be more common in pruned and staked tomatoes than in tomatoes grown on the ground. Blossom-end rot is the formation of black rotted tissue on the tomato fruit opposite the stem. Frequently some of the very first fruit to ripen have blossom-end rot and must be discarded.

Tomato Cages

Sometimes tomatoes are grown in wire cages to keep the plant and fruit up off the ground. A variety intermediate in plant size should be chosen for use in wire cages. Plants that get very tall will grow out the top of the cage and down over the side. If this happens, insect and disease control and picking are more difficult.

The initial cost is greater for wire cages than for stakes, but the labor requirement for growing in cages is much less than for staking and pruning. Labor is required only to place the cages over each plant soon after transplanting and to remove the cages at the end of the harvest season. Wire cages should last at least five years. Cutting the wire and making the cages is a good rainy-day job for your laborers.

To make wire cages, go to your lumberyard and obtain rolls of 6-inch-mesh concrete reinforcing wire made from number-10-gauge wire. The rolls will be 5 feet wide and 150 feet long. One roll will make seventy-four cylindrical cages 14 inches in diameter and 2½ feet tall. Unroll the entire roll and, using wire cutters or a small bolt cutter, cut the roll lengthwise down the center. This produces two rolls of wire 2½ feet wide and 150 feet long.

Next count eight squares from the end of one of those rolls and cut the five wires close to the welds at the wire mesh intersections. This produces a rectangular piece of wire 4 feet long and 2½ feet wide. Bend the 4-foot long section of wire into a cylinder. Bend hooks at the ends of the five wires just previously cut and hook them to the wire on the other end of the 4-foot-long section. You now have a cage 2½ feet high with eight sharp wire legs to push six inches deep into the soil around a small tomato plant. When the cage is pushed into the soil six inches deep, it will be two feet tall and capable of supporting a plant loaded with tomatoes.

Picking tomatoes from cages is much more enjoyable than picking them from the ground. As much as twice as many marketable tomatoes are produced on each plant when cages are used, compared to growing on the ground. The quality of tomatoes grown in cages is very high, and very little fruit cracking or blossom-end rot will occur.

Many good tomato varieties are available. Choose those recommended by your state agricultural college. You may want to grow a few yellow tomatoes and cherry tomatoes. They are becoming more popular. Many customers believe that the yellow varieties are less acidic, but that is generally not true.

Tomatoes are attacked by insects and diseases. Even if you use disease-resistant varieties, your tomatoes will still need to be protected with chemical sprays to avoid complete loss to insects and diseases. Use as long a crop rotation as possible to help avoid soil-borne disease. Potato, pepper, and eggplant have some of the same diseases as tomatoes, and you should keep that in mind when planning your crop rotation system.

Most customers know that vine-ripened tomatoes have the best flavor. Many will pick more dead-ripe

Fig. 21. Tomato fruitworms bore into tomatoes and destroy the fruit. An entire field can be ruined if this worm is present and not controlled. Once it is inside the fruit, the damage has been done, but later fruits can be protected from worms by application of an approved insecticide.

fruit than they can conveniently use before they spoil. Therefore encourage customers to take some tomatoes that are just starting to ripen. They will keep longer after they get them home. Tomatoes picked before they are completely red will go ahead and ripen if they are placed in a cool area not in the refrigerator. Their quality will be as good as that of tomatoes picked from the vine dead-ripe, and they will keep longer.

If you are farming in an area with a long growing season, you can seed your late-summer and fall tomatoes directly in the field. Adjust your planter to plant shallowly several seed per foot of row. Keep the soil moist so that the seed can germinate. Go down the row and thin unwanted plants when they are three or four inches

141

tall. Those left can be grown in cages, pruned and staked, or left on the ground.

SNAP BEAN

Snap bean is a common garden vegetable. It will grow on practically all types of soils. It is also a popular vegetable and should be included in your PYO operation.

Snap beans are a warm-season crop damaged by frost. Planting must be delayed until the danger is over, and snap-bean seed germinates very slowly if soil temperatures are as low as 60° F. If beans are planted in cold, wet soil, a poor stand will result because much of the seed will rot.

New plantings need to be made very week or ten days to have a continuous supply of snap beans. The time from planting to first harvest is usually fifty-five to sixty days. It can be several days less in warmer weather. There are two different types of snap beans, bush and pole. The vines of pole beans must be supported on poles or trellises. A great deal of labor is required to train the beans to the trellises. Pole beans produce over a longer period than bush beans. Therefore less frequent plantings are needed for a continuous supply. Bush beans can be planted in 36-inch rows. Pole beans need about twice that distance between rows to allow space for the trellis system and room for pickers to move between the rows.

Bean roots are shallow. If cultivation is used to destroy weeds, it should not be very deep. The shallow root depth also means that only light applications of irrigation water should be applied. Light, sandy soils should receive one to one-and-half inches of water, and clay soils about three inches of water at each irrigation. If beans are cultivated when the leaves are wet, fungus spores may spread from plant to plant.

There are many snap bean varieties. Cooperative extension service publications in most states indicate the best varieties to grow. The extension services also provide information concerning growing practices and methods to control bean pests.

Snap beans are usually harvested when the seeds are very small and before the pods reach full size. It is best to harvest at that time if the entire pod is going to be eaten. Some customers may want green-shell beans. These are picked when the pods are full size and some pods have started to turn brown. The seed is full sized but still soft at this stage. Dry beans come from completely mature and dry pods. Snap beans are not generally grown for eating as dry beans. Other beans such as Pinto, Navy, Great Northern and Southern peas are grown for dry beans.

Snap beans are usually ready for harvest twelve to seventeen days after the blooms are open. Since the plants bloom over several days, all sizes of beans will be on the plants when the earliest beans are ready to harvest. Thus the beans may be picked several times. Some customers will prefer the smaller beans, and some the more mature beans. If a bush-bean planting is kept watered, it should provide beans to customers for ten days or more before a new planting must be ready for harvest to continue the supply. Pole beans will provide beans over a much longer period of time.

Beans are sensitive to heat during bloom. At high temperatures the flowers will open and then drop off the plant instead of making a bean. Summer temperatures in southern states may be too high to allow season-long production. Bean production will be limited to the spring and fall seasons in other areas where summers are hot. They can be produced during the winters months in the Far South.

Several insects feed on bean leaves and pods. It is

Fig. 22. The snap-bean rows shown here are planted thirty-six inches apart and kept free of weeds. Such clean picking conditions ensure customer satisfaction, and a high-quality crop encourages customers to pick more than they intended.

Fig. 23. A bean and pea sheller will boost sales of garden peas, lima beans, southern peas, and other green shell beans. PYO customers willingly pay extra for this service and buy larger quantities. A machine of this type will shell a bushel in only a few minutes. An employee will be needed to operate the sheller for customers.

important to control them and not allow the bean leaves to be destroyed. Insect damage on snap-bean pods is unsightly and will not be tolerated by customers.

SWEET CORN

Sweet corn is in great demand by PYO customers. Plan on having a continuous supply over as much of your growing season as possible. Since sweet corn is a warm-season crop that is damaged by frost, planting must be delayed until the danger of hard frost is over. Also corn seed germinates slowly at soil temperatures near 50° F and below. Much of the seed planted in cold, wet soil will rot, and a poor stand will result. Yet planting early and taking a chance on frost is worthwhile because early sweet corn is in great demand.

Successive plantings need to be made to have a continuous supply of sweet corn. Varieties differ in the number of days from planting to harvesting. Early varieties take as little as sixty-five days, while main-season and late varieties need from eighty to ninety days to grow. You should use three varieties that are adapted to your area for your first planting, including an early, a midseason, and a late variety. This will provide sweet corn to harvest for about a two-week period. About two weeks after the initial planting, make successive plantings at weekly intervals, using a good midseason variety. A midseason variety is used for all of the later plantings because the ear quality is much better in midseason than in early-season types. Although early-season types are lower in quality, they provide the only early sweet corn available, and customers therefore do not complain.

Your machine planter should be adjusted to plant according to the variety. Early varieties need to be planted about eight inches apart in the row, and you will want midseason varieties about twelve inches apart in the row. Rows thirty-six to forty inches apart produce good yields and leave room for customers to move down the rows to pick the ears. Sweet-corn seed varies a lot in size, but generally eight to twelve pounds of seed are needed to plant an acre. Seed should be planted from one to two inches deep.

There are many good herbicides for use on sweet corn, but if weeds are a problem, cultivation should begin as soon as the sweet corn comes up. Only shallow cultivation should be used to avoid injuring the roots.

Sweet corn uses a lot of nitrogen fertilizer. The common practice is to apply most of the fertilizer before or at the time of planting. When the plants are about one foot tall, additional nitrogen should be side-dressed along the corn row. If heavy rainfall occurs and the sweet corn plants look pale green or yellow, additional nitrogen may be needed.

All ears in a planting will not be ready at the same time, and the inexperienced customer will not know when a sweet corn ear is ready to pick. When a corn ear is ripe, the silk is brown and completely dry; and the ear is plump, indicating well developed kernels. The husk will be very tight around a ripe ear, because the diameter of the ear increases rapidly as it ripens. You may have to teach some customers these signs of ripeness, so that they do not pull down the shuck on all the ears in your field. Sweet corn is of top quality when a kernel is punctured and the interior is milky. The ear is past the best stage for eating if the kernel is doughy. Customers have different preferences, however, and some will want the more mature corn.

Certain varieties will sometimes produce two ears on a stalk. If the stand is thin and each plant has more space, many more stalks will have two ears. Generally,

Fig. 24. Corn earworm is a serious pest of sweet corn in most grow-ing areas. Customers will usually tolerate earworm feeding on the ear tip, but greater amounts of damage cause customer dissatisfaction. Approved insecticides are available to reduce earworm damage.

the top one of the two ears on a stalk will be at the prime stage for picking one or two days ahead of the bottom ear. The ears of modern sweet-corn varieties will be of good quality for four or five days, but the period for picking will be shortened if temperatures are high.

Sweet corn can be grown all season long in northern states. In southern states the hot summer temperatures limit plantings to the spring and fall or in extreme southern areas to the winter. When temperatures are too high, sweet corn will not pollinate, and the ears will have blank spaces where no kernels have formed.

Insects can be a problem on sweet corn, particularly in southern states. The corn earworm feeds at the tip of the ear. To reduce earworm infestation, sprayings should be scheduled from the time the silks appear until they have dried. The rewards for keeping worms out of the corn will be seen at the cash register and by the return of your customers.

147

CABBAGE, BROCCOLI, and CAULIFLOWER

Cabbage, cauliflower, and broccoli are important vegetable crops because of their hardiness to cold. These cole crops can be grown almost anywhere in the United States if proper planting dates are observed. The harvest season for the cole crops can continue after frost has killed warm-season vegetables. In the south cole crops are usually grown in the winter and early spring. In the north they are good fall crops. The best quality is produced during a cool, moist season. When the weather is warm, the harvest period for cole crops will be shortened. For that reason in most regions they are best grown as fall crops so that a longer harvest period is possible.

Cole crops adapt well to almost any type of soil. Fertile, well-drained, sandy loams produce the highest yields for spring crops. During other seasons a heavier loam is better. Cole crops should not be planted in acid soils. Apply lime according to your soil-test recommendation. Because cole crops are heavy feeders on soil nutrients, they need large amounts of fertilizer. Apply fertilizer at planting time and then apply nitrogen along the row at three-week intervals after planting, especially during cool weather and following heavy rains.

Most vegetable growers establish cole crops by using transplants, but seed can be used if desired. Transplants will save two to three weeks growing time. Seeds must be planted two weeks earlier than transplants. Adjust your planter to drop about four or five seeds per row foot in rows three feet apart. If the soil is dry, you must irrigate to obtain seed germination. After the plants grow two or three leaves, thin them with a hoe until they are about twelve inches apart in the row.

Use plants four to five weeks old for transplanting.

If older plants are used, their average head sizes will be smaller than those grown from younger transplants. Transplants may be purchased from a local plant grower, or you can start your own four to five weeks before the desired planting time. Set the plants a little deeper in the field than they were grown as transplants. Water containing fertilizer may be applied as the plants are set. Fall transplants can be grown in open field beds.

Herbicides can be applied and worked into the soil before the transplanting of cole crops. Follow the label instructions. Small weeds can be controlled by shallow cultivation.

Irrigation will be needed to grow a good crop in dry areas. Cole crops grow more rapidly and are of better quality when plenty of water is available. At least one inch of water is required weekly.

Various chemicals will be needed to control disease and insect pests. The chemicals will need to be applied at planting time and throughout the season. Worms can be a real problem in cole crops—be ready to take action to control them. There are several good varieties of cabbage available that are resistant to some diseases.

Cabbage. Cabbage is generally ready to harvest if the heads feel firm when squeezed. Customers will need a knife to harvest this and other cole crops, because they must be cut from the stem. If not harvested when ready, cabbage heads may split open, and in warm weather they will not last long in that state. In cooler weather cabbage heads will keep several weeks in the field. Freezing may occur before harvesting is completed in the fall. Light freezes will injure only the outermost leaves, leaving the rest of the plant still good. Discontinue the harvest when freezing penetrates the heads because rots will develop.

Cauliflower. Cauliflower should be harvested when the head has good size but before it becomes discolored. Customers will usually prefer medium-sized heads. Cauliflower is considered to be of top quality when it is pure white. To obtain pure white cauliflower, the head must be shielded from sunlight. A small cauliflower head is protected from sunlight by small inner leaves that curve up over the head. As the head grows, it becomes exposed to sunlight. Then it is necessary either to "tie" the cauliflower or to break a large outer leaf down over the head. To tie cauliflower, bring several outer leaves together over the head and tie them with a rubber band. Some cauliflower varieties grow large leaves that are upright and shade the head without tying. This self-blanching growth usually occurs in cool weather. In hot weather the self-blanching varieties usually need to be tied like others. The time between harvesting and tying depends on the temperatures. If it is warm, it may be only three or four days from tying to harvest. In cool weather harvest may not be until ten to twelve days after tying.

Broccoli. Broccoli plants are similar in appearance to cauliflowers. For top quality broccoli must be harvested before the flower buds open. After the central head has been cut, small broccoli heads will form in the axil of each leaf. The production of these smaller, secondary heads after the main head has been harvested will provide broccoli for harvest over several weeks.

BELL PEPPER

The sweet pepper has become a very popular raw vegetable crop. Customers who come to your PYO to pick other fruits and vegetables will want to pick some sweet peppers as well, if they are available.

The sweet pepper, which is very sensitive to frost,

Fig. 25. Cabbage sprouts will develop from a cabbage plant after the head is cut. A PYO market can be developed for the sprouts, which are harvested small. They resemble broccoli in appearance, but are very tender and have the milder cabbage flavor.

Fig. 26. To obtain high-quality white cauliflower, the large outer leaves must be tied together over the central head. This keeps sunlight from striking the head and turning the color from white to yellow. Different-colored rubber bands or strings can be used each day that leaves are tied. Customers then can be told that plants with bands or strings of a certain color have heads large enough to harvest and plants with bands of other colors should not be disturbed.

151

basically requires the same climate as the tomato plant. It is a warm-season crop that does well in a long growing season, though in extremely hot weather the plants will drop many blossoms, greatly reducing the yield. They produce the best yields when they are given continuous and fairly consistent amounts of moisture. Some irrigation may be necessary.

Sweet peppers grow best in a sandy loam containing an adequate amount of organic matter. Peppers grow well in soils that are slightly acidic. The soil needs to be plowed in the fall or early spring. A green-manure crop can be planted and plowed under in early spring before discing and harrowing.

Fertilizer needs to be worked into the soil before planting, and the pepper plants need nitrogen fertilizer along the row when they begin setting fruit. Sandier soils require a little more fertilizer than heavier soils.

Transplants should be used to grow peppers. They can be purchased locally, or you can start your own transplants six to eight weeks before the field setting date. Excessive watering should be avoided when growing pepper transplants because of the danger of rotting off. Transplants should be four to eight inches tall when they are set in the field.

Make your pepper rows about three feet apart. Transplants will be set 1½ to 2½ feet apart in the row. They can be set by hand or with a transplanting machine. Set the plants only slightly deeper than they were growing as transplants. After setting, firm the soil around the plants and add one cup of water containing fertilizer to each plant. To avoid root injury, cultivate only lightly to control weeds. Peppers root fairly shallow, and deep cultivation can be very damaging.

Similar diseases affect peppers and tomato plants, and they are treated in much the same manner. Many problems will be avoided by purchasing or growing good-quality plants and by practicing crop rotation. Like other crops, peppers will require some treatment to control insects.

Peppers are normally ready to be harvested when they reach full size and are firm. This is before they turn red or yellow, though many people prefer them turned for canning. Some customers believe green bell peppers become hot when they turn red, but that is not true. Bell peppers usually have a sweeter flavor when they are ripe and red.

Hot peppers are popular in some areas. If enough customers ask for hot peppers, you should plan to grow some. The production of hot peppers is very similar to that of bell peppers. There are many types of hot peppers, but the banana, jalapeño, long chili, and cherry types are usually the most popular.

OKRA

Okra, sometimes called "gumbo," is a popular crop in the southern states. It likes hot weather and grows best under those conditions. The early-maturing small-plant varieties should be grown in northern areas where the season is short.

Okra seed should be planted as soon as the soil is warm. Transplanting will give the crop a quicker start where the growing season is short. Rows should be at least four feet apart, and the plants should be about one foot apart in the row. About four or five pounds of seed are needed to plant an acre. If the plant stand obtained from direct seeding is too thick, the plants should be thinned until they are one foot apart in the row.

Okra will grow in almost any soil. Fertilizer should be applied at the time of planting, and any more nitrogen fertilizer that is needed should be side-dressed along the row and cultivated in. Do not use too much nitrogen because the harvest may be delayed if the plants grow too vigorously. Production will be increased by irrigation if the soil becomes extremely dry.

The long seed pods that develop in the top of the plant are the edible part of okra. The pods increase in size daily and are usually of a harvestable size (3 to 3½ inches long) in five or six days. Okra needs to be picked every day to prevent the pods from becoming too large. Large okra pods are tough and woody and cannot be eaten. As long as okra is continuously harvested, it will produce until frost. If all pods are not harvested from the plant, the flowering will stop, and no new pods will be produced. Therefore do not plant more okra than you expect will be harvested. You may need to send your laborers through the okra planting to remove and discard over-sized pods that customers have not harvested. Removal of these pods will keep the plants producing.

The tender okra pods can be broken or cut off the plant. A knife will be needed for older pods because they become very tough. Okra pods do not have spines, but most varieties have spines on the plant. Customers should be advised to wear gloves and long-sleeved shirts if they plan to harvest okra. Some customers may not want to pick okra. You may want to have some picked for sale to them.

ASPARAGUS

Asparagus is a valuable crop and one of the earliest of the spring vegetables. It has a low ranking as a PYO crop only because customers do not know how to gather it. Yet it can be very profitable even with some waste or destruction by customers. This nutritious vegetable will grow in most of the United States. Production is most successful where freezing temperatures or drought provides for a rest period by stopping growth of the plant.

Asparagus is a perennial crop and may remain productive for fifteen years or more. Spear size may begin to decrease after twelve to fifteen years. A planting should produce 2,000 pounds or more per acre of snapped asparagus during its most productive period. Harvesting asparagus for too long a period weakens the plants by reducing the time for fern growth during the season. Good fern growth is necessary to increase the next year's yields.

Asparagus can be grown on many kinds of soils, but deep loam or sandy loam soils that have good water and air drainage are best. Good production is also possible on heavier soils. It is important that asparagus plants develop an extensive storage root system. Good soil drainage is therefore essential. Asparagus roots can develop to a ten-foot depth in well-drained soils. The crop will do well in moist soils if the water table does not come within four feet of the surface, and it will thrive in soils that have salt content too high for many other crops. Asparagus will tolerate soil conditions that are less than optimum, but yields are likely to be reduced, and life of the planting will be shortened, where soil conditions are bad.

Asparagus occupies the land for many years. It is recommended that the soil be made fertile and free of perennial weeds before crowns are planted, because it is more difficult to improve the soil afterwards. Soil-improving practices must be started at least a year before planting. Limestone should be applied according to soil-test report recommendations if the soil is acidic. Since asparagus grows best in soils well supplied with organic matter, it is good practice to apply animal manure or turn under a green crop before planting. Manure applied to established beds is also beneficial. Specific fertilizer application rates should be determined by soil testing. Most of the required fertilizer should be broadcast and thoroughly mixed with the soil before the crowns are set. Up to fifty pounds of phosphate fertilizer per acre should be placed in the bottom of the planting furrows before setting the crowns.

For a small planting it is best to buy one-year-old crowns from a reliable grower or nurseryman. You may want to grow your own crowns if a large acreage is to be planted. To grow crowns optimally, good seed should be obtained and planted in soil that has never grown asparagus. The soil should be sandy so that the crowns can be easily dug and will be relatively free from adhering soil. Fertilizer should be worked into the soil before the seeding of the crown bed.

Seed should be planted in rows 2 to 2½ feet apart. Ten to twelve seeds should be planted per foot of row. From 1 to 1½ pounds of seed should produce enough crowns to plant one acre of asparagus. Plant the seed one to two inches deep. Asparagus seed is slow to germinate. Three to four weeks are required for the plants to appear above ground.

Crowns should be dug early in the spring before the growth of the buds has begun. Old plant tops should be mowed and removed from the field so that they will not interfere with digging the crowns. A potato digger or common moldboard plow can be used to lift the asparagus crowns from the nursery row. Avoid injury to the crowns when digging and handling. Crowns should be planted as soon as possible after they are dug.

Field uniformity is improved by planting crowns of uniform size. Discard very small or injured crowns and those having many small buds, as they tend to produce a high percentage of small spears. Set the crowns buds up in a wide-bottom furrow five to six inches deep. The planting furrows should be opened with a middle-buster plow just ahead of planting, to assure a loose, moist plant bed. Space the rows four and a half to five feet apart. Space the crowns fourteen to eighteen inches apart in the row. This requires about 6,000 to 8,500 crowns per acre. The distance between rows may be determined by the field equipment used. A large percentage of small spears will be produced if crowns are set too close together. Too much space between plants reduced yields even though the spears are larger.

On small acreages crowns can be carried in a bushel basket and placed by hand in the furrow. Cover the crowns with two to three inches of soil immediately after planting. Work more soil into the furrow as growth progresses, or as is necessary to cover emerging weeds, until the furrow is filled. After the planting of the crowns, the first two years are devoted to developing maximum fern growth in order to build an extensive storage-root system.

After the crowns are set, weeds must be kept under control by cultivation and hand hoeing. Herbicides can be used for weed control in succeeding years. Applications of herbicide can be made in the spring before the first asparagus spears emerge, or one half can be applied before harvest and the other half right after the last harvest when all the spears are removed.

Tillage should be avoided at the end of the harvest season. Discing after harvest destroys all of the spears that remain near the soil surface and delays fern growth. This depletes crown food reserves further and shortens the growing period before frost, thus reducing yields the following season.

157

Fig. 27. Asparagus crowns are planted five to six inches deep and one foot apart in the row. Planting too shallow will cause many pencil-sized spears to be produced. Planting too deep increases spear diameter, but fewer spears are produced, and the yield is much lower.

Irrigation is important to relieve drought stress during the first two seasons after crown planting. An extended dry period early in the fern development after the cutting season may reduce the yield during the following year. Dry weather late in the fern growing season will have little effect on yield.

Asparagus is very deep-rooted and draws water from a large volume of soil. This allows the crop to withstand considerable dry weather. Deep rooting also permits longer intervals between irrigation applications than in other crops. In areas with a lot of rainfall successful asparagus production may be possible in most years without irrigation. In arid areas of the country irrigation is required for high yields.

In the early spring the dead ferns must be chopped down to the level of the soil surface with a brush-hog mower. A shallow discing can follow mowing. Considerable yield loss will result if discing is done too late when spears are near the soil surface, because the early spears that may be destroyed are the largest produced during the season. Do not allow the disc to go deeper than is necessary to control the weeds, especially in sandy soils.

Harvest

Asparagus cannot be harvested in the year of setting crowns because they must be allowed time to grow and develop strong storage-root systems. Harvest for about three weeks during the first year after the crown planting and for about six weeks during the second year after planting. The harvest season can be extended to seven or eight weeks in succeeding years.

The time of asparagus spear emergence depends upon soil temperature. Sandy soils warm earlier than heavy soils and provide earlier production. Harvest will usually begin in late March in warmer areas and in late April or May in colder areas.

There will be some significant losses from customers walking down asparagus rows and snapping very short spears, but the high profits from asparagus will more than make up for those losses. You will need to tell pickers how and which spears are to be harvested. Spears can be cut or snapped when they are six to ten inches above the ground. Always harvest spears before the tips start to "fern out" (open). In the early, cool part of the harvest season spears can grow to nine or ten inches and the heads will still remain tight. These spears will be tender, and the fiber content will be low in their bases. In warmer weather spears will fern out

159

at a shorter height, and more fiber will develop. Therefore, to produce high-quality asparagus, spears should be harvested shorter when temperatures are higher.

The frequency of harvest also depends on the temperature. It may be necessary to harvest daily or even more often when the weather is warm. During cool periods harvesting two or three times a week may be sufficient. Make sure the field is cut or snapped clean with each harvest. This requires close watching. You may have to go over the field yourself to remove unharvested spears that are too tall. Any ferns that are allowed to grow when they should have been harvested may harbor diseases and insects. Such growth also will delay the emergence of new spears.

CUCUMBER

Cucumbers are used in pickling and are also a favorite fresh vegetable in salads. They are a warm-season crop with a short maturing period and can be grown just about anywhere in the United States. Pickling cucumbers are grown from varieties that are different from slicing and salad cucumbers, though the only difference is that slicing cucumbers are about twice as long as pickling cucumbers.

Cucumbers thrive in almost any type of soil, but they prefer a fertile, light loam with good water drainage. Plowing, discing, and harrowing are all necessary to prepare the soil. When those operations are to be done depends largely upon the climate and the drainage of the soil.

Cucumber plants need to have a good supply of fertilizer worked into the soil before planting. Use soil tests to determine how much fertilizer to apply. If heavy rains occur, or if harvesting continues for several weeks, some additional nitrogen fertilizer should be applied along the plant row.

Cucumbers are very sensitive to frost. For that reason many growers make about three successive plantings a week apart, beginning about ten days before the average frost-free date. Cucumber seed is best planted with a planter in a PYO. Rows should be about six feet apart with the seeds three to four inches apart in the row. Cucumber seed is usually planted ½ to 1 inch deep, depending on the condition of the soil. About two pounds of seed will plant one acre.

Transplants can be used to achieve very early cucumber production. Only three to four weeks are required to grow transplants. They should be seeded in peat pots, Jiffy 7s, or fiber blocks to minimize transplant shock. The entire container is planted so that the roots are disturbed very little. Besides the two seed leaves, only one additional leaf should be developed on the plants at the time of transplanting. Larger transplants perform poorly when set in the field.

Cultivation should be shallow to control weeds around cucumbers, and it should not be done too close to the plants. Herbicides can also be used to help control weeds. Normally it will be necessary to use some chemicals to control diseases and insect pests.

It is necessary to have a good supply of bees to pollinate cucumbers. It may be necessary to obtain a bee colony if there are not enough wild bees available. The plants also need a continuous water supply throughout the growing season and particularly when the fruit is setting on the plant. Lack of water can greatly reduce cucumber yield.

The vines should be trained when they begin to run. Move those vines that are growing toward the next row to head down the row. This keeps the vines from filling

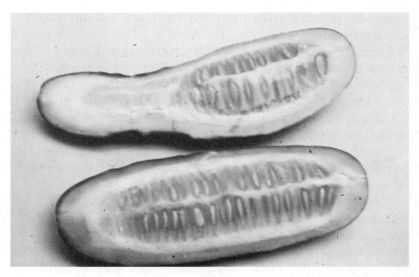

Fig. 28. Cucumbers, melons, squash, and other vine crops need bees to pollinate the female flowers. When bees are not present, fruits fail to develop. When too few bees are present, the fruit develop but may be misshapen. The top cucumber here was not adequately pollinated and therefore developed into a bottle-necked shape. The lower cucumber has a full complement of seeds and a normal shape.

the area between rows where customers will need to walk while picking. If you do not train the vines and keep the rows separated, you will have a mat of vine growth that will be trampled by customers. Yields will be lower, and diseases may be more of a problem.

The first harvest is normally fifty to sixty days after planting. Your customers will prefer medium-sized, dark-green cucumbers that are firm. Cucumbers should be picked every two or three days, and they can be picked daily during their peak period. Cucumbers should not be allowed to become too large and overripen, as that will greatly reduce yields. You may have to send your labor through the cucumber plantings once a week to pick and discard old cucumbers missed by customers. That will keep the vines producing longer. Customers can pick the cucumbers off the vines without any special instructions.

SUMMER SQUASH

Summer squash is becoming an increasingly popular vegetable. It is easy to grow, and only a small area will provide a large amount of squash for your customers. It is a short-season crop that grows in all regions during the summer season. A light freeze will kill the plants.

You should add fertilizer to the soil while preparing the seedbed for summer squash. The rows can be about six feet apart. If they are seeded by hand, four or five seed should be planted in hills about every three feet. If a planter is used, a couple of seeds should be planted in each foot of row. After the plants emerge, the plants from the hand-planted hills should be thinned down to two. If you have sown with a planter, thin so that one plant is left about every three feet in the row. Because the soil must be warm to get good seed germination, wait until the danger of a late spring frost is past before planting.

Summer squash can be transplanted for very early production. They are similar to cucumbers in that only three to four weeks are needed to grow transplants. Use a growing container that can be planted in the field to minimize transplanting shock. Squash plants are ready for transplanting when they have only one leaf developed in addition to the two seed leaves. Larger plants are likely to do poorly.

Additional nitrogen fertilizer should be placed alongside the rows at about the time when harvest begins. Do not be concerned if the first flowers to open drop off

Fig. 29. Almost all the vine crops have male and female flowers. The fruits that are harvested develop from below the petals of the female flower (shown on the right above). Many male flowers (on the left above) are produced for each female flower. Bees and other insects transfer the pollen.

the plant. Squash have male and female flowers, and male flowers always appear first. You can identify the female flowers by looking at the bases. There will be a baby squash located where a female flower attaches to the stem. If the female flowers begin dropping off, it

164

Fig. 30. Bush-type squashes are much easier to manage than vines because they do not cover the areas between the rows as vines do. Customers can walk between the rows without damaging the plants, and cultivation to control weeds is easier.

may mean there are not enough bees in your field. Having a colony of bees working your field will ensure better squash production.

There are many types and varieties of summer squash. Be sure to choose varieties with bushy plants. Those are much easier to manage because the vines will not grow into a jungle. The more popular squash types are the yellow straightneck and yellow crookneck. Also popular is the green summer squash zucchini. All summer squash must be harvested while they are immature, or the skin becomes too tough and the seeds begin to develop. Squash should be no longer than six to eight inches at the time of harvest. Usually only three or four days are required after flowering before the squash are ready to pick. In hot weather it can be even quicker.

If customers do not harvest all the crop, you will need to have your workers go through the planting to pick and discard all oversized squash. The plants will quit producing flowers, and yields will be reduced, if these large fruits are not removed.

Diseases are usually not a great problem in squash, but insects can be. Control measures will probably be necessary to prevent squash bugs from causing too much damage. Control measures should begin early when the bugs are small.

MUSKMELON AND WATERMELON

The muskmelon (more commonly known as cantaloupe) and the watermelon are popular items in the summer. Yet growing them for customers to pick does not work well, because most customers do not know how to select ripe fruit. Muskmelon and watermelon must be picked when ripe if they are to have the sweet flavor enjoyed by most people. You may want to grow some anyway and have your workers pick them. Place them in a shady area near the check-out stand for your customers to buy. Very few customers will leave without purchasing some freshly picked muskmelon and watermelon.

Melons grow best in sandier soils. They can be directly seeded in the field or transplanted. In areas outside the South they should be transplanted, because they require a fairly long growing season. When transplants are used, they should be handled like cucumber or squash: they must be set in the field in growing containers when the plants are small. Whether transplanted or seeded, muskmelons should be spaced about three feet apart in the row. Watermelons can be four to five feet apart.

Melons seeded directly in the field can be hill-planted

by hand and thinned later to two plants per hill. If sown with a planter, they should also, of course, be thinned to two plants per hill. The planter should sow about two or three seeds per foot of row. Since both muskmelon and watermelon are injured by frost, planting must be delayed until frost danger is past and the soil has warmed enough for good seed germination.

Most of the fertilizer should be applied before planting. Additional nitrogen fertilizer will be needed several weeks after planting when the plants begin to make vines. Herbicides and frequent shallow cultivation will help control weeds until the vines begin to run. When the vines have filled the rows, the only way to control weeds is by hoeing or hand pulling.

Irrigation will be needed to produce a good melon crop in dry areas. Irrigation water should not be applied to muskmelons or watermelons when the fruit is nearing full size. Adding water at that time may cause the fruit to crack or split open. Also too much water while the melons are ripening reduces the sweetness and quality of the melons. Everyone likes their melons sweet and flavorful.

Melons suffer from insect and disease problems, and some chemical control of pests will be necessary. If you remember your crop rotation principles, pest problems will be reduced.

A muskmelon should be picked when its stem detaches easily and completely from the fruit. The skin on the muskmelon at that time will have a netting and be changing from green to a lighter green or yellow color. Pick melons before they overripen because they will keep for only a short time after picking before becoming mushy and undesirable.

Selecting watermelons that are ripe is difficult. With experience it can be done quickly, but you will need to

Fig. 31. Watermelon and most other vegetable fruits can develop blossom end rot, which is promoted by improper water and nutrient relations in the plants. Rotten and misshapened fruits should be pruned from the plants when small to improve the quality of the fruits that remain.

train your workers to do it at first. The size of a melon usually is no indication of ripeness, but there are several other qualities to look for. The light-colored area on the underside of each melon, known as the ground spot, usually turns from white to creamy yellow as the watermelon ripens. Therefore roll the melon over slightly to check the ground spot. On the vine near where the watermelon attaches is a tendril—a small, green wiry structure—that intertwines around other leaves and stems. If the melon is ripe, the tendril will usually be dead and drying. A third method is to thump the melon and listen for a dull muffled sound, which is an indication of a ripe melon. A green watermelon will have a more ringing sound when thumped. When all signs indicate that your watermelons are ripe, cut one open and check ripeness before a whole load is picked. All the varieties are a little different in their signs of ripeness, and once you become familiar with them, you will make few mistakes, but should a customer complain about getting a green watermelon, do not argue. Just give the customer the pick of the pile as a replacement and keep him or her happy.

9
Other PYO Opportunities

GROWING ORGANICALLY

There are hundreds of good publications on growing foods organically, and if you are a devotee of organic farming, you probably have all the published help that you will need.

While reading the other parts of the book, you will have noticed that the use of chemicals plays a part in reducing labor; it permits you to achieve high yields and quality by freeing you for other chores. Nonetheless, for many years people have vehemently objected to the use of any chemicals on growing foodstuffs, including chemical fertilizers. Without debating the pros and cons of the matter, suffice it to say that there is an excellent and growing market for foods grown organically.

If you decide to grow part of your crop organically, set aside a special space for it. The area should be upwind from other fields, so that your patrons can feel confident that no chemicals have drifted onto the plants in the organic area.

The prices of your organically grown foods may have to be higher than what you charge for your other crops, because you will have to commit more labor to those fields, for example, in cultivating and hoeing.

Perhaps you want to have a completely organically grown PYO farm. If so, use that to your advantage by advertising your operation as totally organic. Because of the increased interest in foods grown organically, you will probably do a booming business in your first year. If your crops are bountiful, your profits will certainly be great. In fact, they will probably be greater per acre than those of a chemically farmed operation.

UNUSUAL NOVELTY CROPS

In addition to the regular crops suggested in Chapter 8, novelty crops will give you additional income. Examples of such crops are pumpkins (for Halloween), ornamental gourds, decorative corn (for Thanksgiving centerpieces), sunchokes (also known as Jerusalem artichokes), herbs, popcorn, sunflower seeds, garlic, and leeks. Be sure that the new crop will do well in your area. You also need to plan your advertising just right for novelty crops, because timing is very critical. For example, if you are growing pumpkins for Halloween, you need to start advertising about two or three weeks ahead, so that families can plan to come out and get their Halloween pumpkin from you. It is very costly and depressing to have a field full of pumpkins after Halloween when the market for them is past.

GROWING YOUR OWN FOOD

One of the good side benefits of making a living on a farm is that you grow much of your own food. Because you grow small fruits and vegetables in great quantities, you can not only eat the crops while they are in season

171

but also preserve enough for the winter, or off-season, months.[1]

If you would like to have the plans for a good food storage cellar (which could also double as a storm cellar and fallout shelter), the Beltsville Agriculture Research Center of the U.S. Department of Agriculture (Beltsville, Maryland 20705) publishes a set of plans for an excellent facility. Check with your extension agent to see if he has Plan No. 5948, "Storage and Fallout Shelter." If not, write the above address. The Beltsville Center will either send you a copy of the plans or advise you where you can get a copy.

An excellent publication on the how-to of storing foods is Home and Garden Bulletin no. 78, "Storing Perishable Foods in the Home," which can be ordered from the U.S. Department of Agriculture, Office of Governmental and Public Affairs, Washington, D.C. 20250.

When this book was in the initial planning stages, we thought that it would tell not only how to make money growing small fruits and vegetables but also how to be pretty much self-sufficient on an acreage. That approach seemed like a tremendous idea at first, because most of us from a farm background feel that being self-sufficient is a satisfying way of life. Milk, grain, and meat production were to be included among the subject areas. We concluded, however, that, unless your family consumes great quantities of milk, you would be better off financially to buy your milk at the store. First, you would need a place to pasture your cow. Then you would either have to use a certain amount of your land for hay and grain production or buy cow feed. Yet, from a

---

[1] The 1977 issue of the U.S. Department of Agriculture yearbook, entitled *Gardening for Food and Fun,* has an excellent section on home food preservation.

practical viewpoint, the most serious objection to having your own cow is that it is so confining. A cow needs to be milked twice a day, seven days a week. You cannot go off at Christmas time, or any other time, without making arrangements to have the cow milked. If you break your leg or get the flu, the cow still has to be milked. The confinement is the same with other milk-producing animals, such as goats.

Your likes and dislikes will control your meat consumption to a large degree. If you are a beef eater and want to grow and fatten your own beef, you will need special facilities. Such large animals require considerable space, both for pasture and feedlots. Butchering is another problem. Fortunately, some frozen-food locker plants will butcher and cut up your beef before freezing.

The smaller meat-producing animals and birds, such as rabbits, chickens, and quail, are also income producers. Rabbits, for example, have good, tender white meat. Many people who raise them also sell manure from the droppings under the hutches or maintain worm beds under the hutches (worms seem to flourish in this environment). The worms are sold mainly as fish bait. Little space is needed for rabbits, and this is appealing to a small-acreage owner.

Quail and chickens are also good possibilities.[2] Most people enjoy chicken prepared in some way or another, and chickens not only furnish good meat but also eggs. Yet it is hard to compete with the large commercial chicken feeders on production costs. The conversion of feed to meat is excellent for the first few pounds of a chicken. Large chicken growing operations are scien-

[2]J. R. Cain and W. O. Cawley, "Care, Management, and Propagation of Japanese Quail," Publication B-1123, Texas A&M University, and Extension Bulletin E-1069, Cooperative Extension Service, Michigan State University.

tifically operated, and it is very difficult to compete with them costwise, even if you are just raising a few chickens for home consumption.

We made a startling discovery while researching the possibility of raising chickens. Most people who have lived on farms have seen a rooster mount a hen and know that a fertilized egg tastes no different from an unfertilized egg. The process is only important if chicks are wanted because without fertilization eggs will never hatch, no matter how long they are left in the incubator or under the hen. The startling fact is that a rooster can perform his husbandly duties from fifty to a hundred times each day.[3] No wonder that he goes to bed early! He probably crows early in the morning because that is the only time when his battery is charged enough to get out a good crow.

The extent to which you want to engage in farm activities over and above your PYO operation is largely a matter of personal choice. A good PYO operation is very demanding of your time, and unless you particularly enjoy farm animals and birds or want to use them as an attraction for your PYO customers, you probably will be better off without the additional work. The same effort put into your PYO operation will make you more money by far than those incidental activities will.

WAMP'S PYO OPERATION

A gentleman by the name of Okie Wampuskitty has been charged with the crime of creating a public nuisance. Although his story is fictional, it points to some of the problems that can develop for inexperienced PYO

[3] Tankred Doch, *Anatomy of the Chicken and Domestic Birds* (Ames, Iowa: Iowa State Press), p. 99.

operators. It will be quite obvious that Mr. Wampus-kitty was not aware of the management procedures that he needed to operate his PYO operation. Adequate planning is essential, and lack of planning and inexperience are the causes of most of Wampuskitty's problems. The following is an extract of the testimony at his imagined trial.

Defense Attorney: State your name for the record.
Accused: Okie Wampuskitty.
Defense Attorney: Is that your real name?
Accused: Yes sir.
Defense Attorney: People round these parts call you Wamp though, don't they?
Accused: Either Okie or Wamp. Mostly Wamp.
Defense Attorney: Where do you reside?
Accused: Northeast of Muskogee.
Defense Attorney: Muskogee, Oklahoma?
Accused: Yes sir.
Defense Attorney: Now Wamp, you've been charged with a crime and you've heard the witnesses against you and you deny the charges, do you not?
Accused: Well, I guess I clogged up the road in front of my place, but I didn't mean to cause nobody trouble.
Defense Attorney: Tell the jury what happened.
Accused: My wife and I decided to try to raise strawberries. We was looking forward to the harvest period, which lasted about three weeks. I was going to hire anybody who would work during this three-week period. The harvest was supposed to produce most of our annual income.

My wife was pregnant and was having trouble carrying the baby. She was to rest as much as possible. I knew that I would be having trouble trying to harvest without her.

With my wife down most of the time, I was having to do a lot of the housework like cooking, getting the kids ready for school, and so on. Anyway, I didn't seem to have time to find hands to harvest the strawberries. So I decided

175

to place an ad in the big city newspaper to let the people know they could come to my place and pick their own strawberries. It was the only way I could figure to get the strawberries picked.

I decided to start on Saturday because my older kids would be there to help me. The strawberries would be ready for the first picking on that Saturday. My strawberries were ready about a week before the Arkansas berries, for some reason. I ran the ad for a week before that Saturday because I wanted to be sure all the strawberries were picked.

Saturday morning came. I was eating breakfast when a car horn honked. I looked outside and saw a car with three ladies in it wanting to know if this was the place where you could pick strawberries. I said yes and went out to show them where to go. When I left the house, I never got to come back in until after dark.

The ladies left to pick strawberries, and they were followed by two more cars. These were followed by five more cars. Then I lost count. You won't believe how many cars came that day. My yard filled up right off. People were parking on both sides of the highway which passed by my place. The cars stacked up for a mile in both directions. There was room for only one car to go down the road and then the middle got blocked. People couldn't get to my place, and those who were there couldn't get out. Some of the people became mad. My neighbors were mad because they couldn't get in or out of their places. My wife was mad because there was a long line waiting to use our bathroom. I finally told the men and boys to go to the bathroom behind the chicken house and the ladies to go inside the barn. A lady with a French accent told me she was not going to use a pissoir. I was afraid to ask her what she meant. People were lined up to use my telephone. Either one of my customers or my neighbors called the sheriff. When the sheriff came, I'm sure the sight shocked him. He couldn't find the owners of the cars. A lot of them were in my strawberry field.

A lot of the people brought their dogs. One of the little female dogs was in heat. The other dogs were either chasing

that little female or fighting. Things got out of hand—way
out of hand. The sheriff went back to town and got a portable public address system. When he returned, he cut my
fence and crossed over into my fields so he could come down
to the house in his car. When I explained to him what was
going on, he couldn't believe it. He got on his public address system and told everybody to return to their cars.
Those people with their containers full of strawberries started
for their cars. I tried to get the sheriff to announce that
the people should pay for the strawberries before they left.
He told me to quit bothering him because he was going to
unclog the road. The ladies who had just arrived protested
having to leave with their empty containers. The sheriff
didn't listen to their protests.

There were quite a few who would not use the chicken
house or the barn for a bathroom, and there was a line to
our bathroom. One little boy peed on our new rug, and I'm
still hearing about that. The sheriff couldn't budge those
in the bathroom-line who appeared to be in pain.

One by one, the cars began to move out. The only problem was that cars continued to come from the big city. The
sheriff had to place two deputies on the highway leading
to my place telling them to go home. Those people were
mad. I told the sheriff that the next day might be about
the same or worse, since it would be Sunday and the city
folks might be out in great numbers. He told me I had
better leave the fence down that he had cut and let the
customers park in the field. I did. The sheriff was kind
enough to station one deputy at my place the next day to
direct traffic, but I was told to put up signs because this
would be the only time the deputy would be stationed there.
The sheriff told me he was filing a charge against me for
causing that big mess.

People are still talking about that day. I hope most of the
folks are over their mad spell. One lady isn't, because she
wrote me a letter threatening to sue me, claiming I got her
water system out of order because there were no bathroom
facilities that day. I hope the poor thing doesn't get the word

she is pregnant. No doubt, she would point the finger at me, since I have been blamed for everything that happened that day.

After that first day, I began making some money.

Defense Attorney: Pass the witness.

Court: You may cross-examine.

Prosecutor: No questions.

Court: Call your next witness.

Defense Attorney: Your Honor, the defense has no other witnesses except Mr. Wampuskitty's wife, and her testimony would be cumulative. The defense rests.

Court: Rebuttal witnesses?

Prosecutor: None, your Honor.

Court: The Court will be in recess until three o'clock.

The Court reconvenes at three o'clock.

Court: Ladies and gentlemen of the jury, I have concluded that, as a matter of law, there is insufficient evidence on which to base a conviction. You are therefore excused from further jury duty until tomorrow morning at 9:00 a.m. I want to thank each of you for your attentiveness during the trial. Please leave the courtroom quickly and quietly.

Court: Bailiff, assist the members of the jury with their exit.

The members of the jury departed.

Court: Gentlemen, I have concluded that based upon the testimony heard against Mr. Wampuskitty, that as a matter of law, there is insufficient evidence to support a conviction and I am therefore dismissing the charges. If Mr. Wampuskitty is guilty of anything, it is of mismanagement and of course that is not a crime. Mr. Wampuskitty, you are free to go. The Court is adjourned until 9 a.m. tomorrow morning.

ANALYSIS OF WAMP'S OPERATION

Although Wamp's operation and trial are fictional, his

initial experience as a PYO operator is not far from that of anyone who is just beginning, especially for strawberry and sweet-corn growers. The first day can be very hectic.

There are some good features about Wamp's operation. He evidently had good technical knowledge of how to grow strawberries. He probably had irrigation, or else there was just the proper amount of rain at the right time and there were no late freezes. One thing certainly was going for him: his location. He must have had a strategic location to draw such crowds.

Many of the unfortunate happenings at Wamp's place could have been avoided if he had planned ahead. First, his advertising was so effective that it brought too many customers on the weekend. Strawberries will deteriorate fast if they are not harvested when ripe. Wamp may not have had the option of starting his picking before the weekend, but if that option is available, it is better to start on a weekday. The crowds then will not be nearly as large. If Wamp had no option in the matter, he should not have advertised for so long during the week before the harvest was to start.

When you do start the harvest, it is absolutely necessary to have a parking area that can take overflow cars. There was no way that Wamp could accommodate the cars that came to his place. He also had no toilet facilities, which are a must for an operation of any size. We cannot know about his check-out procedure because, as it developed, he did not have the opportunity to check people out—the sheriff chased everybody off the premises. Undoubtedly, he did not have any insurance for protection against possible lawsuits.

It is not unusual for people to enter the pick-your-own business the way Wamp did, except that there usually is no courtroom activity. Many people decide to let customers come and pick the harvest without any thought of what is involved. Wamp would have been much better prepared if he had followed the procedures for retailing fruit and vegetables at the growing site that we have set out in the preceding chapters.

Bibliography

This bibliography is a fairly comprehensive list of publications relating to PYO farming. Some of the publications concern only the operations of roadside stands, but they may still be of interest, since a roadside market can be a part of any good PYO operation.

Most of the publications are excellent. Many will give you important insights into ongoing PYO operations. Those that are out of print have been included in case they are reprinted. The publications of an older vintage contain principles that are still valid today.

The publications are grouped under the names of the states, because you undoubtedly will want to secure those by experts in your state. The District of Columbia is included because the United States Department of Agriculture publications originate there. All USDA publications are available from the Superintendent of Documents, United States Government Printing Office, Washington, D.C.

Following the listing by state and district is a selection of books under the heading "Production Information" on the growing of fruits and vegetables. Like many of the publications in the state listings, these books and pamphlets can be found in major libraries or ordered from the publishers, if they are not in local book stores. Your county cooperative extension office is also a major source of publications. Extension office publications are usually free, but some have a small fee. All the publications of your state agricultural university are usually available at the extension office. These publications will be particularly valuable to you because they are written by experts in your state agricultural university and are specific to the production of crops in your area. United States Department of Agriculture publications are written in general terms that are applicable to all states.

ALABAMA

Fabian, M. S. "Pick-Your-Own Marketing: An Alternative for Producers and Consumers." In *Proceeding of a Symposium on Marketing Alternatives for Small Fruit and Vegetable Farmers, February 19-25, 1979,* edited by W. J. Free (Tennessee Valley Authority, Muscle Shoals, Ala.), p. 58.

Free, W. Joe. "The Farmers' Marketing Needs in the South." In *Proceeding of a Symposium on Marketing Alternatives for Small Fruit and Vegetable Farmers, February 19-25, 1979,* edited by W. J. Free (Tennessee Valley Authority, Muscle Shoals, Ala.), p. 10.

German, Carl L., and Mary Deckers. "Roadside Marketing." In *Proceeding of a Symposium on Marketing Alternatives for Small Fruit and Vegetable Farmers, February 19-25, 1979,* edited by W. J. Free (Tennessee Valley Authority, Muscle Shoals, Ala.), p. 41.

How, R. Brian. "Overview of Direct Marketing Alternatives." In *Proceeding of a Symposium on Marketing Alternatives for Small Fruit and Vegetable Farmers, February 19-25, 1979,* edited by W. J. Free (Tennessee Valley Authority, Muscle Shoals, Ala.), p. 18.

Mizelle, William O., Jr. "Overview of Commercial Marketing Alternatives for Fresh Fruits and Vegetables." In *Proceeding of a Symposium on Marketing Alternatives for Small Fruit and Vegetable Farmers, February 19-25, 1979,* edited by W. J. Free (Tennessee Valley Authority, Muscle Shoals, Ala.), p. 32.

BIBLIOGRAPHY

Williams, J. Louis. *Answers to Your Questions About Direct Marketing of Fruits and Vegetables*. Circular no. 152, Auburn, Ala.: Auburn University Cooperative Extension Service, 1980.
————. *Consumer Survey Results*. Auburn, Ala.: Auburn University Cooperative Extension Service, 1980.

ARKANSAS

Lambert, R. E. *Roadside Markets for Vegetables and Fruits*. Fayetteville, Ark.: Division of Agriculture, University of Arkansas, 1975.
Lambert, R. E., and C. C. Schaller. *Pick-Your-Own Fruits and Vegetables*. Fayetteville, Ark.: Division of Agriculture, University of Arkansas, 1978.

CALIFORNIA

Amorocho, G., and L. Garoyan. *Direct Marketing Opportunities: Applications for Sonoma County*. Berkeley, Calif.: University of California Cooperative Extension Service, 1973.
Cothern, J., and M. B. Hall. *Direct Marketing: The Results So Far*. Berkeley, Calif.: University of California Cooperative Extension Service, 1977.

COLORADO

Cloman, L. Ruth. *Direct Producer-to-Consumer Marketing: A Case Study*. Fort Collins, Colo.: Colorado State University Cooperative Extension Service, 1979.

DELAWARE

Ginder, R. G. *Advantages, Problems and Suggestions for Pick-Your-Own Marketing of Horticultural Products*. Newark, Del.: University of Delaware Cooperative Extension Service, n.d.
Ginder, R. G., and H. H. Hoecker. *Management of Pick-Your-Own Marketing Operations*. Newark, Del.: Northeast Extension Marketing Committee, University of Delaware Cooperative Extension Service, 1975.

GEORGIA

Brown, E. E., and R. L. Jordan. *Roadside Marketing in Georgia*. Research Report, no. 254. Athens, Ga.: University of Georgia Department of Agricultural Economics, 1977.

Spivey, C. D., and Paul Colditz. *Growing for You: Pick-Your-Own the Profit Making Way.* Circular no. 643. Athens, Ga.: University of Georgia Cooperative Extension Service, 1972.

ILLINOIS

Courter, J. W. *Pick-Your-Own Marketing of Fruits and Vegetables.* Urbana, Ill.: Department of Horticulture, University of Illinois, 1979.

Courter, J. W.; Roberta Archer; C. M. Sabota; and R. E. Westgren. "Direct Marketing of Fruits and Vegetables: A Fresh Approach." *Illinois Research* (University of Illinois Agricultural Experiment Station, Urbana, Ill.) 21, no. 1 (Winter, 1979).

Courter, J. W. *Estimating the Trade Area and Potential Sales for a Pick-Your-Own Strawberry Farm.* Urbana, Ill.: Department of Horticulture, University of Illinois, 1982.

Courter, J. W., and C. C. Zych. "Survey of Pick-Your-Own Strawberry Customers." In *Better Farming in Illinois: A Report from Horticulture.* Urbana, Ill.: University of Illinois Cooperative Extension Service, 1969.

Sabota, C. M., and J. W. Courter. *Net Weights and Processed Yields of Fruits and Vegetables in Common Retail Units.* Urbana, Ill.: University of Illinois Department of Horticulture, 1979.

Uchtmann, Donald L. *Liability and Insurance for U-Pick Operations.* Urbana, Ill.: University of Illinois Department of Horticulture, 1979.

INDIANA

Blakely, Ransom A. *Packaging and Displays for Roadside Marketing.* West Lafayette, Ind.: Purdue University Department of Horticulture, 1971.

———. *Parking Area Design for Farm Markets.* West Lafayette, Ind.: Purdue University Department of Horticulture, 1972.

———. *Preparation and Use of Signs in Roadside Marketing.* West Lafayette, Ind.: Purdue University Department of Horticulture, 1972.

———. *Pricing in Perspective.* West Lafayette, Ind.: Purdue University Department of Horticulture, 1975.

———. *Roadside Market Layout.* West Lafayette, Ind.: Purdue University Department of Horticulture, 1970.

———. *Some Conclusions on Sign Design.* West Lafayette, Ind.:

BIBLIOGRAPHY

Purdue University Department of Horticulture, 1971.
——. *Suggestions for a Beginning in Roadside Marketing.* West Lafayette, Ind.: Purdue University Department of Horticulture, 1971.
——. *U-Pick Management.* West Lafayette, Ind.: Purdue University Department of Horticulture, 1970.
Kurschling, Patrick J., and Glenn H. Sullivan. *Small-Farm Costs and Returns: Pick-Your-Own Vegetables.* Bulletin no. 223. West Lafayette, Ind.: Purdue University Agricultural Experiment Station, 1979.

KANSAS

Morrison, Frank, and Alice King. *The Apple Farm Market Case Study.* Manhattan, Kans.: Kansas State University Department of Horticulture, 1979.
——. *A Roadside Farm Market: A Case Study for Improved Practices.* Manhattan, Kans.: Kansas State University Cooperative Extension Service, 1979.

KENTUCKY

Davis, Hubert W. *Roadside Marketing in Kentucky.* Lexington, Ky.: University of Kentucky Agricultural and Home Economics Extension Service, 1974.

LOUISIANA

Roy, Ewell P.; Don Leary; and Jerry M. Law. *Customer Evaluation of Farmers' Markets in Louisiana.* Research Reports no. 534 (March, 1977) and no. 516 (June, 1978). Baton Rouge, La.: Louisiana State University Department of Agricultural Economics and Agribusiness.

MAINE

Metzger, Homer B., and Wilfred H. Erhardt. *Marketing Fresh Fruits and Vegetables Through Roadside Stands and Pick-Your-Own Operations in Maine, 1974.* Bulletin no. 724. Orono, Maine: University of Maine Life Sciences and Agricultural Experiment Station, 1976.
—— et al. *Marketing Fresh Vegetables Through Roadside Stands.* Bulletin no. 710. Orono, Maine: University of Maine Life Sciences and Agricultural Experiment Station, 1975.

MARYLAND

Hoecker, Harold H. *Pick-Your-Own Fruits and Vegetables in Maryland.* College Park, Md.: University of Maryland Cooperative Extension Service, 1978.
——. *Roadside Markets in Maryland.* College Park, Md.: University of Maryland Cooperative Extension Service, 1978.

MASSACHUSETTS

Ames, William S.; Donald R. Marian; and Robert L. Christenson. *An Economic Evaluation of Direct Marketing: A Massachusetts Case Study.* Amherst, Mass.: University of Massachusetts Department of Food and Economics, 1979.
Bahn, Henry M. *Direct Marketing Options for New England Producers.* Amherst, Mass.: University of Massachusetts Cooperative Extension Service, 1980.

MICHIGAN

Antle, Glenn G. *Roadside Marketing for Beginners.* East Lansing, Mich.: Michigan State University Cooperative Extension Service, 1978.
——. *Pick-Your-Own: Another Marketing Option for Michigan Fruit and Vegetable Growers.* East Lansing, Mich.: Michigan State University Cooperative Extension Service, 1978.
Kelsey, M. P., and Hugh Price. *For Pick-Your-Own Operations: Computing Production Costs of Fruits and Vegetables.* Bulletin no. E941. East Lansing, Mich.: Michigan State University Cooperative Extension Service, 1979.
Stachwick, George. *Successful Roadside Marketing: A Manager's Manual.* East Lansing, Mich.: Michigan State University Cooperative Extension Service.

MISSISSIPPI

Pullin, A. T. *Roadside Markets: Fruits and Vegetables.* Jackson, Miss.: Mississippi State University Cooperative Extension Service, 1978.

MINNESOTA

Kinnucan, Henry, and Ben Seraver. *Consumers Find Alternative*

Food Sources in Minnesota. Saint Paul, Minn.: University of Minnesota Department of Agricultural Economics, 1977.

NEBRASKA

Jannsen, D. E., and D. H. Steinegger. "Roadside Marketing." *University of Nebraska Farm, Ranch, Home Quarterly* (Lincoln, Neb.), Summer, 1977.

NEW HAMPSHIRE

Dalton, M. M., and R. A. Andrews. *Marketing Agricultural Products in New Hampshire: Sales Patterns in Roadside Markets.* Durham, N.H.: University of New Hampshire Institute of Natural and Environmental Resources, 1979.

NEW JERSEY

Fabian, Morris S. *Opportunities in Roadside Marketing.* New Brunswick, N.J.: Rutgers University Cook College Cooperative Extension Service, 1972.

――――. *Tips for Retail Farm Market Salespersons.* New Brunswick, N.J.: Rutgers University Cook College Cooperative Extension Service, 1975.

――――. *Tips for Retail Farm Market Salesperson's Supervisors.* New Brunswick, N.J.: Rutgers University Cook College Cooperative Extension Service, 1975.

―――― and Mark G. Robson. *Guidelines for Marketing Pick-Your-Own Produce in New Jersey.* Bulletin no. 428. New Brunswick, N.J.: Rutgers University Cook College Cooperative Extension Service, 1979.

NEW YORK

Blakeley, R. A. *A Comparison of Alternative U-Pick Marketing Strategies for Strawberries.* Ithaca, N.Y.: Cornell University Agricultural Economics Department, 1978.

How, R. Brian. *Prospects for Direct to Consumer Sales of Fresh Fruits and Vegetables.* Ithaca, N.Y.: Cornell University Agricultural Economics Department, 1974.

Stuhlmiller, E. M., and R. Brian How. *Selected Characteristics of Direct Marketing Businesses: Six Counties in New York, 1976.* Ithaca, N.Y.: Cornell University Agricultural Economics Department, 1978.

NORTH CAROLINA

Proctor, E. A. *Some Pointers on Roadside and Pick-Your-Own Marketing.* Folder no. 319. Raleigh, N.C.: North Carolina State University Cooperative Extension Service, 1974.

Williamson, Lionel. *Factors to Consider in Direct Marketing Through Mobile Roadside Stands.* Greensboro, N.C.: North Carolina Agricultural Extension Service, A & T State University, 1979.

OHIO

Bennett, Thomas A.; M. E. Cravens; and J. E. Jeffries. *Retail Farm Marketing in Northeast Ohio.* Wooster, Ohio: Ohio State University Agricultural Research and Development Center, 1967.

Cravens, M. E. *Roadside Marketing by Ohio Farmers.* Wooster, Ohio: Ohio State University Agricultural Research and Development Center, 1972.

Hungate, Lois S., and E. Watkins. *Characteristics of Customers of Pick-Your-Own Operations.* Columbus, Ohio: Ohio State University Department of Agricultural Economics and Rural Sociology, 1978.

Ohio Roadside Marketing Conference *Proceedings,* 1966-68, 1971, 1972, 1977-79. Columbus, Ohio: Ohio State University Department of Agricultural Economics and Rural Sociology.

Watkins, Ed. *Developing Merchandising Plans for Ohio Roadside Markets.* Columbus, Ohio: Ohio State University Department of Agricultural Economics and Rural Sociology, 1978.

———. "50 Ways to Promote Markets." *American Vegetable Grower* (Meister Publishing Company, Willoughby, Ohio), June, 1979, p. 48.

———. *Financial Planning for Roadside Marketing.* Columbus, Ohio: Ohio State University Cooperative Extension Service, 1978.

———. "How Customers See Your Roadside Market." *American Vegetable Grower* (Meister Publishing Company, Willoughby, Ohio), June, 1978, p. 8.

———. *Keeping Food Wholesome in Ohio Roadside Markets.* Columbus, Ohio: Ohio State University Department of Agricultural Economics and Rural Sociology, 1978.

———. *Ohio Roadside Market Management and Marketing Practices.* Columbus, Ohio: Ohio State University Department of Agricultural Economics and Rural Sociology, 1978.

BIBLIOGRAPHY

_____. "Why Direct Marketing?" *American Vegetable Grower* (Meister Publishing Company, Willoughby, Ohio), June, 1979, p. 8.

OREGON

Fetters, T. A., and H. A. Meier. *Direct Farm Marketing Practices and Activities in Oregon.* Special Report no. 570. Corvallis, Ore.: Oregon State University Cooperative Extension Service, 1980.

PENNSYLVANIA

Kerstetter, Raymond J. *Direct Marketing and the Pennsylvania Department of Agriculture.* Harrisburg, Pa.: Pennsylvania Department of Agriculture, 1976.

Toothman, J. S. *A Checklist for Developing Your Market Sales and Operating Plans.* University Park, Pa.: Pennsylvania State University Department of Agricultural Economics, 1974.

_____. *Guide for Handling Customer Contacts.* University Park, Pa.: Pennsylvania State University Cooperative Extension Service, 1974.

_____. *Market Conditions Checklist.* University Park, Pa.: Pennsylvania State University Cooperative Extension Service, 1974.

_____. *Measures of Retail Merchandising and Operating Performance.* University Park, Pa.: Pennsylvania State University Cooperative Extension Service, 1974.

_____. *An Outline for Informing Employees of Basic Policies and Rules Pertaining to their Employment.* University Park, Pa.: Pennsylvania State University Cooperative Extension Service, 1974.

_____. *Steps in Hiring, Training and Supervising Roadside Market Clerks.* University Park, Pa.: Pennsylvania State University Cooperative Extension Service, 1974.

_____. *U-Pick Farm Retailing Guidelines.* University Park, Pa.: Pennsylvania State University Cooperative Extension Service, 1974.

SOUTH CAROLINA

Harrelson, William L. *The Roadside Market Incentive Program in South Carolina.* Columbia, S.C.: South Carolina Department of Agriculture, 1972.

McLeod, Revel. *South Carolina Roadside Market News.* Columbia, S.C.: South Carolina Department of Agriculture, 1973.

TENNESSEE

Bird, J. J. *Pick-Your-Own Fruits and Vegetables.* Knoxville, Tenn.: University of Tennessee Cooperative Extension Service, 1970.

Sams, D. W.; D. W. Lockwood; and A. D. Rutledge. *Pick-Your-Own Fruits and Vegetables.* Knoxville, Tenn.: University of Tennessee Cooperative Extension Service, 1980.

VIRGINIA

Bell, James B. *Conventional Fruit and Vegetable Marketing Programs for Small Farmers, Southern Small Farm Management Workshop Proceedings.* Blacksburg, Va.: Virginia Polytechnic Institute Department of Agricultural Economics.

———. *Direct Marketing in Virginia.* Blacksburg, Va.: Virginia Polytechnic Institute Department of Agricultural Economics, 1978.

VERMONT

Pelsue, Neil H. *Consumers at Vermont Fruit and Vegetable Roadside Stands: Part I.* Research Report no. 2. Burlington, Vt.: University of Vermont Agricultural Experiment Station, 1980.

WASHINGTON, D.C.

Cate, H. A., and J. W. Courter. *Pick-Your-Own Catches On. Extension Service Review* (U.S. Department of Agriculture), June, 1970.

Jones, Judith L.; Richard B. Smith; and Charles R. Handy. "Direct Marketing: Consumers' View." *National Food Review* (U.S. Department of Agriculture) no. 3 (June 1978). An article on 1978 study of consumers' use and awareness of direct marketing outlets.

Lindstrom, H. R. *Farmer-to-Consumer Marketing.* ESCS-01. Economics, Statistics, and Cooperative Service, U.S. Department of Agriculture, 1978.

Watkins, Ed. *Direct Marketing Bibliography: 1970-1980.* Science and Education Administration, U.S. Department of Agriculture, 1980.

BIBLIOGRAPHY

———. *Direct Marketing Opportunities.* Science and Education Administration, U.S. Department of Agriculture, 1979.
———. *Displaying in a Farm Market.* Science and Education Administration, U.S. Department of Agriculture, 1980.
———. *Farm Market Management with Emphasis on Pricing.* Science and Education Administration, U.S. Department of Agriculture, 1979.
———. *The Potential for Direct Marketing.* Science and Education Administration, U.S. Department of Agriculture, 1980.

WEST VIRGINIA

Haines, C. William. "Experiences with Pick-Your-Own Fruit." In *Proceedings* of National Peach Council 31st Annual Convention, Atlantic City, N.J., February 20-23, 1972, Martinsburg, W.Va.: National Peach Council.
Stadelbacher, G. L. "Retail Farm Marketing: Pick-Your-Own Fruit." In *Proceedings* of National Peach Council 31st Annual Convention, Atlantic City, N.J., February 20-23, 1972. Martinsburg, W.Va.: National Peach Council.

WISCONSIN

Klingheil, George C. *Pick-Your-Own Strawberries: The Ten P's to Profit.* Madison, Wis.: University of Wisconsin Cooperative Extension Service.
Reed, Robert H. *Roadside Marketing.* Madison, Wis.: University of Wisconsin Cooperative Extension Service, 1975.
——— and J. A. Glaze. *Pick-Your-Own Marketing.* Madison, Wis.: University of Wisconsin Department of Agricultural Journalism, 1978.

PRODUCTION INFORMATION

Childers, Norman F. *Modern Fruit Science.* New Brunswick, N.J.: Horticultural Publications, Rutgers University, 1973.
Gardening for Food and Fun. USDA Yearbook of Agriculture. Washington, D.C.: U.S. Government Printing Office, 1977.
Knott, James Edward. *Vegetable Growing.* Philadelphia: Lea and Febiger, 1958.
Lorenz, Oscar A., and Donald N. Maynard. *Knott's Handbook for Vegetable Growers.* New York: John Wiley and Sons, 1980.
Ourecky, Donald K. "The Strawberry Grower's Handbook." *Amer-*

Shoemaker, James S. *Small Fruit Culture*. Westport, Conn.: AVI Publishing Company, Inc., 1975.

Splittsloesser, W. E. *Vegetable Growing Handbook*. Westport, Conn.: AVI Publishing Company, Inc., 1979.

Teskey, Benjamin J. E., and James S. Shoemaker. *Tree Fruit Production*. Westport, Conn.: AVI Publishing Company, Inc., 1978.

Thompson, Homer C. and William C. Kelly. *Vegetable Crops*. New York: McGraw-Hill Book Company, Inc., 1957.

Ware, George W., and J. P. McCollum. *Producing Vegetable Crops*. Danville, Ill.: Interstate Printers and Publishers, Inc., 1975.

Index

Pick-Your-Own Farming,
designed by Sandy See, was set in
Century Text and a version of Helvetica
by the University of Oklahoma Press and
printed offset on 60-pound Husky, with
presswork by Cushing-Malloy, Inc. and
binding by John H. Dekker & Sons.